12/19 (42) LAD12/18 unique

D1013012

Cosmic Numbers

ALSO BY JAMES D. STEIN

How Math Can Save Your Life
The Right Decision
How Math Explains the World

COSMIC NUMBERS

The Numbers That Define Our Universe

James D. Stein

BASIC BOOKS

A MEMBER OF THE PERSEUS BOOKS GROUP

New York

Library of Congress Cataloging-in-Publication Data
Stein, James D., 1941–
  Cosmic numbers : the numbers that define our universe / James D. Stein.
    p. cm.
  Includes bibliographical references and index.
  ISBN 978-0-465-02198-7 (hardback) — ISBN 978-0-465-02788-0 (e-book)
1. Cosmological constants. 2. Cosmology. I. Title.
  QB991.C658S74 2011
  530.11—dc22
                        2011008539

10 9 8 7 6 5 4 3 2 1

TO BILL BADE,

with great appreciation for all your help

CONTENTS

PREFACE

Something I had never experienced before happened while I was writing this book.

I'd written several previous books, but I'm not sufficiently well established as an author to simply write a book and have a reputable publisher bring it to market. Like most prospective authors, I have to write a proposal, which consists of an outline of the book, its potential market, and a couple of sample chapters. My agent then shops it to various publishers, and—with luck—someone offers to publish it.

I've always been fascinated by numbers, and it occurred to me that the discovery of the numbers that are the heart of this book—the Cosmic Numbers, if you will—would make for a fascinating book. There are very few new ideas under the Sun, and this idea had occurred to other authors as well. Martin Rees had written a book called *Just Six Numbers* (a few of which are in this book) describing the six numbers that he felt lie at the heart of cosmology, but there were other numbers that I felt also deserved to have their stories told. So I wrote up the outline for the book, as well as an introduction and a sample chapter on Absolute Zero. To my great joy, not only did Basic Books, the leading publisher of scientific trade books, decide to publish it, but T. J. Kelleher, whom I knew to be a terrific editor because I had worked with him previously on *How Math Explains the World*, agreed to work on it.

I knew T. J. to be a great editor because, among other reasons, when we were working on the previous book, he spent a great deal of time structuring the sequential organization of the chapters. This greatly added to the flow and readability of the book; his choice was not the one I proposed but it unquestionably worked out better. I did not think that organization would be a similar problem with this book, as the cosmic numbers it discusses belong to three branches of physical science: physics, chemistry, and astronomy. I initially saw the book as being organized along those lines, and started work on the obvious first chapter—the gravitational constant.

What made the process of writing this book remarkable was that each chapter seemed to presage the next, organizing itself by the historical development of science rather than by grouping the chapters by discipline. After a few chapters, I realized that I was writing an outline of the history of science as exemplified by the numbers that I had decided to use. It's not a complete history of science by any means; the life sciences are nonexistent and the development stops somewhere in the middle of the twentieth century. Nonetheless, if you give this book to someone who knows absolutely nothing about science (which unfortunately describes a large segment of the American public), by the time they finish it, they'll have a very good idea of what has happened in the major physical sciences. It's history by the numbers—though not in the conventional sense of the word.

Several other things worth mentioning happened while I was writing this book. While doing the background reading that this book required, I had an opportunity to read the biographies of several of the scientists whose contributions appear herein. I don't know which impressed me more—the quality of the writing or the scholarship displayed in the thorough researching of the individuals involved. Some of these books are listed in back, but the ones that absolutely blew me away were *The Master of Light*, the exquisitely detailed story of Albert Michelson (written by his daughter); the short but extraordinary *Ludwig Boltzmann* (written by Englebert Broda), a book that made you wish you had the opportunity to spend one hour in Boltzmann's company; and *Chandra* (by Kameshwar Wali), a description of the professor who awes—and to some extent, terrifies—students, but who is universally admired and loved by his colleagues.

Four people made substantial contributions to enable this book to be written. Quite simply, T. J. Kelleher edits like no one else I have ever encountered. Even when some of my favorite passages have been excised, it is almost invariably with complete justification, and the book is nearly always better as a result. I also noticed a disconnect between T. J.'s style and mine in the first chapter or so that he revised, but after that, when I read the revised chapter, it almost seemed to me as if I had written it completely! I have no idea how he does it; I can only write in my own style—and my guess is that every one of T. J.'s authors would attest to this ability. It helps to have an editor who not only notices the flaws in your presentations, but when he remedies them, it seems like you have written the material. Finally, T. J. has a love for science and mathematics that one rarely finds in anyone other than a scientist or a mathematician. I've only encountered one other such individual—and that person was my father, coincidentally also a Harvard man, as T. J. is.

I owe my writing career to my agent, Jodie Rhodes. These are difficult times for authors, as publishers are often unwilling to take risks, and it must be extremely difficult for agents to encounter rejection and still be willing to stand by their authors and fight for their rights in an environment where sales are difficult to obtain. Well, it may be difficult for other agents, but Jodie has pitched for me and battled for me under conditions that can only be described as arduous and disheartening. While I think I'm a passable writer, it is necessary to find an editor and publisher who share this opinion, and Jodie has extensive experience that has enabled her to match me with editors and publishers who appreciate my efforts. Possibly other agents could have achieved this— but I doubt it, and I have no idea what I will do if she retires.

The third person is one of the most remarkable students I have ever had the pleasure to teach. Sometime in the 1980s, Dave McKay enrolled in an upper-division course in mathematical analysis that I was teaching. I have counted Dave as a friend and colleague ever since, and this book has benefited enormously from the fact that Dave, a fellow faculty member at California State University, Long Beach, has not only become an extremely adept instructor of mathematics, but an equally proficient instructor of physics as well. I've always loved physics, but it's been as one who worships the object of his passion from afar, as I have never understood the great ideas of physics with the same clarity that I understand some of the great ideas of mathematics. Dave does—because he has been willing to put in twenty-five years of simply studying physics with an eye toward understanding it the way mathematicians understand mathematics.

The reader of this book will notice a large number of calculations, because not only is this book about the cosmic numbers that define the universe, it is about numbers themselves—the universal language, as Galileo called mathematics. Most of the calculations in this book require no more than very basic algebra, geometry, or maybe a little trigonometry, but often there is a physical theory in the background that underlies these calculations. The rationale for the physical theory lies outside the scope of this book, but most introductory physics texts contain all the equations and formulas I have used.

Last—but not least—is my wife Linda. I'm not overly fond of the song "You Are the Sunshine of My Life"—the tune isn't so great and the lyrics are a little on the mawkish side—but it's a good description of Linda. She doesn't write the books, but she does the stuff that makes it a lot easier for me to do so. Just as some people complain that math makes their brains go all fuzzy, contracts have the same effect on me— I can't read more than a paragraph or so, and Linda has the tenacity to go through them with a fine-toothed comb. Of course, that's an added bonus thrown in with being the sunshine of my life.

By the time this book is published, I'll be seventy years old, and really have only two regrets, both concerning my parents. They never got to read any of my books, and they never met Linda. I think they would have enjoyed both experiences.

CHAPTER 1

THE
GRAVITATIONAL
CONSTANT

I t is impossible for me to fully grasp life in the seventeenth century, during which Isaac Newton spent the greater part of his life. It was a world of alchemy rather than chemistry, a world without many of the simple things that make life bearable (at least for me): no toilet paper or toothpaste, telephones or televisions. But it was a world of books and newspapers, of letters and journals (the seventeenth-century version of blogs), and as a result we know almost as much about Isaac Newton as we would if he had walked around with a GPS tracking device affixed to his ankle—assuming the device had been attached around the year 1664.

Newton, however, was born in 1642, which leaves a pretty large gap in any biography of the man. From what we do know, it seems clear that, unlike the cases of such prodigies as Mozart or the mathematician Carl Friedrich Gauss, he did not do anything in his youth to presage his future greatness. What we do know is that his mother wanted him to become a farmer. Fortunately for us, Newton displayed a complete lack of interest in farming, but it took the joint effort of the headmaster of his school (who seems to have been the only individual to have recognized Newton's potential) and Newton's uncle to persuade his mother to send Isaac to Trinity College in Cambridge. He entered his "safety school" in 1661. It was one of the most successful plan Bs in history.

His early years at college are also not well documented, either by himself or his contemporaries. His journal has records of presumed highlights ("At the Taverne twice") and lowlights ("Lost at cards twice"), but there isn't a hint of the genius that was to emerge. Things began to take off in 1664, when, as he noted in his "Waste Book" of jottings, he began a serious study of mathematics. Prior to that, Newton's knowledge of mathematics seems to have been at the level of a contemporary high school sophomore; the evidence is that he was comfortable with arithmetic but his knowledge of algebra, geometry, and trigonometry might not have been enough to produce an impressive score on the SAT. Newton brought himself up to speed by either purchasing or borrowing the texts that were state-of-the-art for mathematics at the time. From Oughtred's *Clavis Mathematicae*[1] ("The Key to Mathematics"), he learned the power and flexibility of algebra—which would lead to his discovery of the general binomial theorem. From Wallis's *Opera Mathematica*[2] ("Mathematical Works") he gained his initial insights into what would later become his signature mathematical accomplishment—the development of the infinitesimal calculus. Newton relied on a Latin translation by Schooten of Descartes's *Géométrie*[3] to rectify his geometrical deficiencies.

He would receive his bachelor's degree in 1665, the year of the last great outbreak of the bubonic plague in England. Plague spreads under crowded, unsanitary conditions—and this was sufficiently well recognized that the court of King Charles II departed London for Oxfordshire, and Cambridge University closed. Isaac Newton chose to return to his childhood home in Woolsthorpe—and spent the next eighteen months "minding Mathematicks & Philosophy."[4] In so doing, he remade the world.

The Development of the Theory of Gravitation

As fundamental as Newton's contributions to mathematics were, it is nonetheless for his contributions to science that he is most remembered, for it is by advances in science that much of the progress in the human condition is made. He made formidable contributions to optics, but of course it is for his work on mechanics and gravitation, and secondarily the scientific method of theory and experimentation, that he is held in such esteem.

The first enunciation of a scientific theory is almost invariably not the simplest. Innovators such as Newton are not generally concerned with presenting material so that it can be understood by the widest possible audience; they are more interested in having it become accepted by their peers, and then building upon it. Such is the case with Newton's *Philosophiæ Naturalis Principia Mathematica*[5] ("Mathematical Principles of Natural Philosophy," generally referred to as the *Principia*); I have opened it on occasion and have resolved to read it when I retire (adding to my list of still-unkept resolutions). The style of Newton's *Principia* resembles a standard geometry text—axioms, theorems, lemmas, proofs—and many of the conclusions are, in fact, geometrical. This is not surprising, because one of the key achievements of the work, which in part is a description of Newton's theory of gravitation, was its ability to explain Kepler's three laws of motion, all of which are geometrical. Kepler's first law states that planets have elliptical orbits around the Sun, with the Sun at one focus of the ellipse. The second law states that an imaginary line drawn from the center of the Sun to the center of the planet will sweep out equal areas in equal intervals of time. And the third law states that the ratio of the squares of the periods of any two planets is equal to the ratio of the cubes of their average distances from the Sun.

These laws were not just the insights of a brilliant geometer working from a few premises; they were also empirical—the result of a lifetime of data-gathering and model-fitting, building on the data painstakingly amassed by Tycho Brahe, an eccentric Danish nobleman with an interest in astronomy. Brahe had been impressed by Kepler's early work, and invited Kepler to visit him near Prague, where Brahe was constructing a new observatory. Kepler would become Brahe's intellectual heir.

At the time, the Copernican revolution was gaining steam, and Kepler attempted to fit Brahe's excellent data to Copernicus's model of the solar system, which held that the planets move in uniform circles around the Sun. Indeed, Kepler's initial model of the orbits of the planets had an extra wrinkle, as he thought they corresponded to geometrical properties of the five regular Platonic solids—tetrahedron, cube, octahedron, dodecahedron, and icosahedron, with four, six, eight, twelve, and twenty faces respectively.

At any rate, Kepler tried to fit the data he had to circles. Fortunately, Brahe had just obtained highly accurate observations of the planet Mars—and the orbit of Mars deviates substantially from a circle. Had Brahe just finished observations of Venus, whose orbit is almost perfectly circular, it is not clear when—or even if—Kepler would have been able to arrive at his first law.

Kepler's achievement in discovering the first law is a testament to his real intellectual rigor, and the second and third laws to his substantial mathematical ability. Finding the area of the elliptical sectors needed for the second law is a task considerably beyond basic Euclidean geometry, and recognizing the power relationship inherent in the third law also requires considerable mathematical *savoir faire*. Nevertheless, Kepler spent years formulating and checking the second and third laws. Through all of this, Kepler was beset by numerous personal and political problems—he lost both his wives and his favorite son to illnesses, and his refusal to convert to Catholicism limited his potential for employment. On top of this, Kepler had to provide the legal defense when his mother was accused of witchcraft, a charge that in those days could result in death by torture. The charges were based solely on rumor, however—not surprising, because to the best of my knowledge, there haven't been a lot of authenticated cases of witchcraft, either then or now, and Kepler was able to obtain her acquittal.

Kepler's accomplishments more than justified his epitaph:
"I measured the skies, now the shadows I measure,
Sky-bound was the mind, earthbound the body now rests."[6]

A Question of Velocity

A straightforward, nonquantitative conclusion from Kepler's first and second laws is that planets move at different speeds at different locations in their orbit. An ellipse is a stretched-out circle, with a profile like a blimp's, and has two axes of symmetry, long and short. If the ellipse in question is a planetary orbit, the Sun will be found on the long axis close to the ellipse. Now, imagine that a planet travels a small distance from just above the long axis near the Sun to just below the long axis near the Sun. We can approximate the area it sweeps by using the area of an isosceles triangle (although the planet's path is curved, over small distances it's reasonable to treat it as a straight line perpendicular to the long axis). The height of the triangle is the distance of the Sun to the ellipse along the long axis, less than half the length of the long axis because we positioned the Sun on the long axis near the ellipse. It is clear that if the planet is traveling at the same speed at all times, it will traverse the same distance along its path when it is close to the Sun or at the symmetrical location on its orbit far away from the Sun. Suppose that the planet always travels at the same velocity. If the planet travels the same small distance from just above the long axis far away from the Sun to just below the long axis far away from the Sun, the area that it sweeps out according to Kepler's second law can again be approximated by a triangle with a base having the same length as the base of the triangle near the Sun. This time, however, the height of the triangle—the distance from the Sun along the long axis to the ellipse, is more than half the length of the long axis, and so the two triangles have different areas. If Kepler's first and second laws are to hold, the planet cannot be traveling at the same velocity when it is near the Sun as when it is far away from the Sun.

Newton's work on calculus would be invaluable for explaining how that happens. One of the great insights afforded by calculus is the means to define constantly changing quantities—such as the speed of a planet, or a car—at any given moment. So, for example, imagine I drove one afternoon from Los Angeles to San Diego, a distance of 120 miles, in three hours. Simple arithmetic tells me my average velocity during the trip was 40 miles per hour, but it can't tell me how fast I was traveling when I went through the open stretch just before Interstate 405 turns into Interstate 5, or how slowly I was traveling in the traffic jam near Mission Viejo. To determine how fast my car was traveling at 2 p.m., we need to look at the collection of average speeds of my car over successively shorter time intervals at that time. The average velocity of the car computed over a time interval of one second is a more accurate approximation to the actual speed of the car at the start of the interval than is the average velocity of the car computed over a time interval of one minute—because there is a lot more time for the car to change its velocity in a one-minute span than in a one-second span. If we were to measure that average velocity over even shorter intervals—say an interval of .001 second—it is extremely close to the exact speed of the car at the start of that interval, assuming, of course, I haven't rear-ended a truck during that .001 second.

Newton's *Principia* recognizes not only this, but states a method for computing the instantaneous velocity at any time by means of what calculus students learn as the difference quotient method, which involves taking limits of averages. He also presages the difficulty many calculus students have with this.

"I chose rather to reduce the demonstrations of the following propositions to the first and last sums and ratios of nascent and evanescent quantities, that is, to the limits of those sums and ratios; and so to premise, as short as I could, the demonstrations of those limits. For hereby the same thing is performed as by the method of indivisibles; and now those principles being demonstrated, we may use them with greater safety. Therefore if hereafter I should happen to consider quantities as made up of particles, or should use little curved lines for right ones, I would not be understood to mean indivisibles, but evanescent divisible quantities; not the sums and ratios of determinate parts, but always the limits of sums and ratios; and that the force of such demonstrations always depends on the method laid down in the foregoing Lemmas."[7]

I have a pretty good knowledge of calculus, but wading through Newton's explanation in the preceding paragraph is not easy for me, and I would think it would be almost impossible for a twenty-first-century student to learn very much from his book, whether calculus or his theory of gravitation.

Big *G* and Little *g*

At the core of Newton's work on gravitation, there are actually two constants: the universal constant *G* that is described in the *Principia*, and the local acceleration *g* at the surface of Earth due to the force of gravity. Little g, as it is often called, is relatively easy to measure, at least if we are willing to settle for an approximation valid to two or three decimal places—all we have to do is find a vacuum (so as to eliminate air resistance), drop stuff, and measure how far it falls and how long it takes to fall. It was Galileo who originally realized that the distance that stuff falls is proportional to the square of how long it has been falling, and it was one of the many consequences of Newton's law of gravitation—and a simple problem in first-semester calculus— to show that the distance d that an object falls in time t is $d = \frac{1}{2} gt^2$. Little g was determined fairly easily to be approximately 32 feet per second per second. It's easier to think of this as "32 feet per second"— pause—"per second"; every second that an object falls under Earth's gravitational influence increases its velocity by 32 feet per second. Things fall much more slowly on the Moon, as the astronauts demonstrated—even Wile E. Coyote has time on the Moon to get out from under the falling anvil. Little g, therefore, is a local constant.

Big G, on the other hand, is universal, but there is a relation between G and g, as you might expect. One of Newton's achievements was to show that the gravitational force of a sphere acts as if all the mass were concentrated at the center. Therefore, the gravitational force on an object of mass m exerted by Earth (whose mass we will denote by M and whose radius by R) is given in two ways: by $F = GmM / R^2$ by the law of gravitation, and by $F = mg$ by Newton's second law of motion. Equating these two expressions, we see that the term m cancels on both sides of the equation, and that $g = GM / R^2$. The value of R was known (approximately) to the ancient Greeks—but in order to determine G to any accuracy, it is necessary to know the value of M, and no inroads were made on this problem until well after Newton died.

In fact, there was no real interest in determining G for almost two centuries, because nothing of what the scientists of the day wanted to learn required the knowledge of the value of G. Much of what was done in astronomy—and indeed, what is still being done—involved using ratios. That's not so surprising, for the equality of ratios enables many a practical computation, and had done so long before the *Principia*. Ratios show up early in arithmetic. (If two eggs are needed for a batch of cookies that will feed three children, how many eggs will be needed for enough cookies to feed twelve children?) They show up again in geometry, when we use the equality of the ratios of corresponding sides of similar triangles to measure the height of an unclimbable tree—or a distant mountain. Both of these uses of ratios—arithmetical and geometric—are of immense practical importance in the physical sciences, as well as in everyday life. Without the proper number of eggs, you're not going to be happy with the way the cookies crumble.

Newton could derive Kepler's third law—the ratio of the squares of the periods of any two planets is equal to the ratio of the cubes of their average distances from the Sun—from his law of gravitation. Astronomers could then use these ratios, combined with distance from Earth to the Sun (which had been calculated by Giovanni Cassini more than a decade prior to the publication of the *Principia*)[8] and the periods of the planets to compute the average distance of a planet to the Sun. There simply wasn't a need to know the gravitational constant—and so nobody bothered to compute it until an experiment that took place at the end of the eighteenth century enabled its value to be known.

The Cavendish Experiment

Most of the great scientists leave more than just the record of their theories or their experiments, they leave memories such as participation at conferences and professional or personal exchanges with other scientists. But just like our everyday world, the world of science has its loners—and among them was Henry Cavendish, one of the great experimental scientists of the eighteenth century.

We know that Cavendish was born in France in 1731 to Lord Charles Cavendish and Lady Anne Grey, and benefited from a huge inheritance. He dropped out of Cambridge after three years without obtaining a degree, but this did not prove to be any sort of impediment to his scientific career. His personal life had its own share of impediments, however, as social occasions and personal relationships seemed to be very difficult for him. He was painfully shy around women, even going so far as to communicate with the female household servants via written notes and building special staircases and entrances for them in his houses so that he would not have to run into them. Social engagements with

Cavendish apparently weren't worthy of a journal entry, either in his or anyone else's. The only record of his public appearances seems to be when he would attend a scientific conference.

The noted physician and author Oliver Sacks has suggested that Cavendish had Asperger's syndrome, which resembles autism in that those afflicted by it have difficulty interacting with others and display repetitive behavior. But repetitive behavior, or at least the willingness to repeat the same thing over and over again, is just what you need if you're going to be an experimental scientist, and Cavendish made notable contributions to both chemistry and the study of electricity. Among these was his analysis of the components of air. He discovered that air was approximately 20 percent "flammable air" (oxygen), and 80 percent nitrogen—although he also noted that approximately 1 percent of air consisted of gases other than these two; it would be a century before the existence of argon as an element and its presence in the atmosphere was confirmed. He also did pioneering work in the study of "inflammable air" (hydrogen), and is responsible for discovering that hydrogen and oxygen are the chemical components of water; he came extremely close to the correct H_2O formula.[9]

His contributions to the study of electricity were also noteworthy—he was the first to study dielectric materials (those which do not conduct electricity) as part of a study of electricity, and he was the first to distinguish between electrical charge and voltage. He was also the first to study the conduction of electricity in water, prompted by the reports that some fish were able to produce electric shocks—he actually modeled a fish from leather and wood soaked in salt water, gave it imitation electricity-producing organs, and demonstrated that a fish could indeed produce an electric shock. Although Cavendish did very little in the way of publication, he did record his notes, and it is a measure of the esteem in which he was held by the British scientific community that no less a scientist than James Clerk Maxwell took it upon himself to go over Cavendish's notes and publish them to make sure that Cavendish posthumously received the credit that he deserved.

The experiment for which Cavendish is best known—and which is referred to nowadays as "*the* Cavendish experiment"—was the one that first determined the density of Earth. This was Cavendish's purpose, but his experiment has often been called "weighing Earth," for once Earth's average density has been determined, its weight can be determined with good accuracy simply by multiplying that density by Earth's volume. In fact, so well known was this experiment that for years thereafter his neighbors would describe the building where it was conducted as the place where Earth was weighed. Considering that his public appearances bordered on the nonexistent, it can safely be said that Cavendish was truly a scientist whose reputation preceded him.

The experiment, employing what is known as a torsion balance, was a masterpiece of ingenuity. Two large heavy balls are fixed in place, and two small balls are placed at opposite ends of a very thin wire, resembling a small dumbbell. This is suspended at the midpoint of the wire. The gravitational attraction between the heavy balls and the smaller balls causes the smaller balls to rotate very slightly (the amount of rotation would be a lot larger if magnets were used to produce the deflection rather than gravity, an indication of how much stronger magnetism is than gravity). The amount of rotation can be measured, and can be used to compute the average density of Earth—or its mass. So accurate was Cavendish's apparatus that his estimate was not improved for a century.

Hidden in Cavendish's data was a way to compute the gravitational constant—but since nobody really cared about the gravitational constant at the time, nobody bothered to compute it. Today's physicists would take Cavendish's data and compute the gravitational constant in a relatively straightforward fashion.

Let M be the mass of one of the larger balls, and let L be the length of the thin dumbbell-shaped wire. Let θ be the angle through which the wire rotates, and let r be the distance between the centers of the small and large balls after the wire has rotated. Finally, let T be the natural oscillation period of the balance (akin to the period of a pendulum). The following formula for the gravitational constant G is obtained by equating two forces on the smaller ball: the gravitational force from the larger ball and the restoring force from the rotating wire (the gravitational force pulls the small ball toward the larger one; the restoring force is the same type of force that a stretched spring exhibits as it tries to return to its unstretched position). Modern physicists would obtain the following:

$$G = 2\pi^2 L r^2 \theta \,/\, MT^2$$

Cavendish actually used the same quantities to compute the average density of Earth, which he obtained by using Newton's second law of motion, equating the net force mg on the small ball with the gravitational force $GmM_E \,/\, r_E^2$, where M_E and r_E are, respectively, the mass and the radius of Earth. We could do this as well. Denoting the average density of Earth by ρ, since its volume is $4\pi r_E^3 \,/\, 3$, we obtain $\rho = 3g \,/\, (4\pi G r_E)$. Cavendish actually computed the density as 5.448 grams per cubic centimeter—but in communicating this result he made an uncharacteristic error of leaving out a 4 and reported it as 5.48 grams per cubic centimeter.

We tend to think of anything prior to the era in which we were born as relatively primitive, and the end of the eighteenth century—when the cause of disease was unknown and horseback the fastest mode of transportation available—verges on the Paleolithic. Nonetheless, Cavendish's experiment was incredibly accurate, and thanks to the fantastic collection of resources currently available on the Internet, you can actually read Cavendish's own words on this experiment.[10]

He may not have had today's resources, but he took a tremendous amount of care in planning and executing the experiment. He also was intellectually honest—he begins his communication for the *Philosophical Transactions of the Royal Society of London* with "MANY years ago, the late Rev. John Michell, of this society, contrived a method of determining the density of the Earth, by rendering sensible the attraction of small quantities of matter; but, as he was engaged in other pursuits, he did not complete the apparatus till a short time before his death, and did not live to make any experiments with it. After his death, the apparatus came to the Rev. Francis John Hyde Wollaston, Jacksonian Professor at Cambridge, who, not having conveniences for making experiments with it, in the manner he could wish, was so good as to give it to me."[11] Michell is also known as the individual who first postulated the existence of a black hole. It seems to me that history is really short-changing Michell here; it was *his* idea and *his* equipment, and this might be a good time to start calling this the Michell-Cavendish experiment.

And what better time to start doing this than now?

Modern science recognizes the importance of determining the values of the basic constants. The Committee on Data for Science and Technology (CODATA) periodically collects the most recent values for the basic constants. The latest updating of *G* that I could find is in the 2006 CODATA[12] report, and the section on the gravitational constant begins, "The HUST (Huazhong University of Science and Technology) group . . . determines *G* by the time-of-swing method using a high-Q torsion pendulum with two horizontal, 6.25 kg stainless steel cylindrical source masses labeled A and B positioned on either side of the test mass . . . "![13] More than two centuries after Michell and Cavendish, with all the advances in technology that have come since then, the method they suggested is still cutting-edge. Six of the eight measurements involved in determining the gravitational constant involve torsion balances.

Why We Need to Know *G* As Accurately As Possible

It's not just one of those nerdy things of interest only to the Henry Cavendishes of the world.

The gravitational constant is basic to the universe; its existence has been known perhaps longer than any other fundamental constant, and yet its value is known only to five significant digits—less accurately than any of the constants discussed in this book. This is due in large measure to the extreme weakness of the gravitational force when compared with the other forces (the electromagnetic force, the strong force, and the weak force). There are potential advances in measurement on the horizon. The 2006 CODATA section on the gravitational constant mentions that experiments are under way to determine the gravitational constant using atom interferometry, which analyzes wave patterns. However, there may be another approach that uses existing data—a lot of it.

If an object is circling Earth in a spherical orbit of radius r, it can be shown that the orbital period T, the time for it to go once around Earth, is given by $T = 2\pi r^{3/2} / (GM)^{1/2}$, where M is the mass of Earth. If one considers r, G, and M as unknowns, given sufficient objects in circular orbits, I would think that it would be possible to measure T and r to a high degree of accuracy for each of them, and given any collection of two different objects, there would be two equations for G and M. These could be solved for all possible pairs of objects in circular orbits, and the results for G and M could then be subjected to statistical analysis. Even if the orbits are not circular, there is an equation for the orbital period in terms of the orbital parameters—and there is a lot of debris currently orbiting Earth.

Maybe we can't measure accurately enough, maybe our computers are not yet sufficiently powerful to perform this analysis, and maybe there is a reason to rule this out on the basis of a theorem in statistics, but even that would be worth knowing. NASA maintains a huge database of all the debris that's in orbit, and if I were a data miner, I'd sure consider unpacking my data pick and my data shovel to go looking for data gold in them thar hills.

But why should we care? One reason is that this could pose a problem for future spaceflights, especially journeys to the stars if we are ever capable of making them. I'd hate to run out of fuel before getting to Alpha Centauri just because we didn't know G to enough decimal places. However, a more pressing reason for seeking a more accurate value of G is that it would enable us to determine more accurately the future positions of comets and asteroids that might pose a threat to Earth. Forewarned might enable us to be forearmed.

CHAPTER 2

THE
SPEED OF
LIGHT

My interest in math and science makes me look for math and science in unexpected places—in particular, in the lyrics to some of my favorite songs. When Jim Morrison of the Doors wrote, "The crystal ship is being filled, a thousand girls, a thousand thrills, a million ways to spend your time,"[1] my first reaction (other than the pleasure of listening to the song) was to wonder how familiar Morrison might have been with basic combinatorics, which is essentially the science of counting. Because he was right: if you participate in each one of a thousand thrills with each one of a thousand girls, in addition to being completely exhausted, you will indeed have found a million ways to spend your time.

Some years later, Bob Seger wrote (in "Night Moves"[2]), "I woke last night to the sound of thunder. *How far off?* I sat and wondered." I knew he was from Detroit, but didn't he either join the Boy Scouts or pay attention in science class? You don't have to sit and wonder how far off the thunder is, you simply have to count one-thousand-one, one-thousand-two . . . from the moment you see the lightning flash until you hear the sound of thunder. To be fair to Seger, as copyeditor Sarah Van Bonn points out, he might not have seen the lightning flash if he really had awoken to the sound of thunder. Nonetheless, counting in this fashion comes pretty close to one count per second, and the speed of sound is approximately one mile every five seconds, so if you get to one-thousand-five when you hear the thunder, you know that the lightning hit about a mile away. (In the Boy Scouts, we also learned what to do if the lightning flash and the thunderclap are very close together—drop to the ground and curl yourself up in a ball. Perhaps you worry about these things more if you live in the suburbs than if you live in the central city.)

Galileo knew something like this as well. I'm not sure when people first became aware that sound travels at a speed that could be fairly easily measured, but by the seventeenth century, thanks to the proliferation of cannons, the lag between the sight and sound of an explosion was well known. In his *Dialogue Concerning Two New Sciences*,[3] Galileo proposed using a simple analogue of this phenomenon to measure the speed of light. Two men would stand facing each other, each holding a light. They would both cover the light with their hands; then the first would uncover his light, and when the second saw this light he would uncover his. Galileo recognized that this would be impractical at short distances, but that with the aid of the recently invented telescope, this could be done over substantial distances. Unfortunately for Galileo, who actually tried to perform this experiment, the distances involved were totally inadequate to enable this method to work. Light moves so rapidly that it traversed the longest distance over which Galileo conducted the experiment in less than a ten-thousandth of a second—a duration that could not be measured in Galileo's era. As a result, Galileo concluded that the speed of light was either infinite or extremely rapid.

Nevertheless, Galileo's idea was sound—find a distance over which light takes some measurable period of time to travel, record the time, measure the distance, and use the fact that when something travels at a constant velocity, the velocity at which it travels is equal to the distance traveled divided by the time it takes to do so. Despite being unable to perform this computation himself, Galileo made one of the most important observations in the history of science, which not only revolutionized man's view of the universe, but made possible the first determination of the speed of light.

The Moons of Jupiter

The invention of the telescope is generally credited to the Dutch lens maker Hans Lippershey, who applied for a patent on the device and made it generally available in 1608. The most common use was by merchants, who would scan the distant ocean to see whether they could spot incoming vessels. On January 7, 1610, Galileo trained a telescope toward Jupiter, and observed "three fixed stars, totally invisible by their smallness,"[4] close to Jupiter and collinear with it. Later observations showed these objects were moving in a way that would not have been possible had they indeed been stars. On January 10, he noticed that one of the "stars" had disappeared, which he correctly attributed to its having moved to a position where Jupiter blocked its light. Within a few days he had concluded that the "stars" were actually orbiting Jupiter.

This revolutionary discovery was to shake the world, because if there were celestial objects orbiting a body other than the Earth, our planet could not be the center of the universe as the dominant theology of the time required. This discovery, famously, put Galileo into direct opposition with the Catholic Church. Perhaps it is a measure of progress that heretical scientific theories have met with less severe consequences as time has passed—Giordano Bruno was burned at the stake in 1600 for espousing a cosmology in which the Sun was just one of countless stars, whereas Galileo was merely confined to house arrest in 1633, and John Thomas Scopes would only have to pay a $100 fine in 1925 for teaching evolution in a Tennessee classroom.

Galileo's discovery of the moons of Jupiter also provided Ole Rømer, a Danish astronomer, with a way to estimate the speed of light. The Italian Giovanni Cassini had made observations of the eclipses of the Jovian moons, and had noticed that the interval between the eclipses changed, shortening as the distance between Earth and Jupiter decreased, and lengthening as that distance increased. Cassini reached the conclusion that this was due to the fact that light took longer to reach Earth from a greater distance, and actually announced this in 1676 to a meeting of the French Academy of Sciences.[5] He concluded that it took light between 10 and 11 minutes to traverse the distance from the Earth to the Sun. By that time, this distance was reasonably well established, and using 10½ minutes for the time it took light to travel this distance results in a value for the speed of light of 93,000,000 miles / (60 × 10½) seconds, or about 147,000 miles per second (which is about 80 percent of the actual value we know today). This appeared to have been a throwaway result for Cassini, who turned his attention to other matters. Rømer made a series of observations of the eclipses of the moon Io lasting eight years, and published his observations. Just as Cavendish would have data more than a century later to determine the gravitational constant but didn't bother to do so, Rømer had the data for determining the speed of light, but he also didn't bother to calculate it. Like Cavendish, neither Cassini nor Rømer actually computed the value of that fundamental constant, but I guess the theory is that if you do enough of the spadework, you deserve the credit, and Rømer is generally credited with the first determination of the speed of light.

Rømer's estimate was somewhat improved by James Bradley, an English astronomer, some fifty years later. Bradley was relying on the same astronomical measurements available to Rømer, but a conceptual breakthrough enabled him to do a better job of it. Bradley, while out sailing one day, was watching the pennant flutter on the mast. Regardless of how he steered the boat, the pennant maintained a constant direction because the wind did. It occurred to Bradley that light acted the same way as the wind, while the Earth, like the boat, moved. This gave Bradley a better estimate of how long it took light to travel from the Sun to Earth, and consequently a better value for the speed of light— specifically he thought it would take light 8.2 minutes to reach our planet from the Sun, which is about 1.2 percent faster than it actually does. It would take more than a century for the technology to improve to a point where it was possible to bring the technique for measuring the speed of light back down to Earth.

It's All Done with Mirrors

In the middle of the nineteenth century, two French physicists, using similar approaches to the problem, came up with methods for measuring the speed of light that would provide the impetus for a deep revolution in physics. The first technique was devised by Armand-Hippolyte-Louis Fizeau, and was based on an idea of an earlier French physicist, Dominique François Arago, whose eyesight was so bad he could not carry out his experiment. Fizeau positioned two mirrors opposite one another separated by a distance of 8,633 meters—about $5^1/_3$ miles. He placed a rapidly rotating toothed wheel between the two mirrors and shone a light beam between the teeth of the wheel. He then adjusted the speed of rotation so that the returning light beam hit the gap between the next two teeth of the wheel. Since technology had advanced to the point that the speed of rotation could be kept constant, which wasn't possible without machines to control this, the time could be calculated merely by knowing the rotation rate of the wheel and the number of teeth in the wheel. Fizeau's measurement was about 5 percent too high, but it was nonetheless a considerable improvement over Rømer's estimates. Although Bradley's estimate was better, the technological innovation of this experiment would pave the way for even more accurate determinations of the speed of light. In both science and mathematics (as I tell my students), sometimes the method of computation is more important than the results of the computation.

Also tackling this problem was Fizeau's friend Jean Bernard Léon Foucault, who used a similar approach to Fizeau. This was not surprising, since the two had been good friends since their college days, and had actually considered a joint project for measuring the speed of light, but after an argument, they separated and decided to pursue the problem independently. Foucault's technique also involved two mirrors set up some distance from each other, but instead of passing light through a cogwheel, he reflected it off a rotating mirror, powered by a steam engine he had constructed himself. This was directed toward the second mirror, and would then reflect off it and hit the initial mirror, which had by now rotated slightly. Fizeau had used the rotation of the toothed wheel to clock the time for light to make the round trip; Foucault computed this time by measuring the angle by which the returning light beam was deflected.

This apparatus was devised in part prior to the argument that broke up the Fizeau-Foucault partnership, and Foucault also used the rotating mirror technique to show that light traveled more slowly through water than it did through air. Just as Cavendish acknowledged the role that Michell had played in devising the torsion balance, Foucault acknowledged Fizeau. Well, *almost*. Here are Foucault's words:

"I did not invent the spinning mirror, nor the achromatic lens, nor the network, nor the micrometer, but I have had the good fortune to be able to put these instruments, devised by other scientists, together in such a way that I have solved a problem which was posed twelve years ago."[6]

It seems as if the argument with Fizeau still rankled, and although Foucault was intellectually honest enough to own up to the developments leading up to his experiments, he felt that, because he included other devices along with Fizeau's spinning mirror, he needn't acknowledge Fizeau personally.

Foucault's experimental apparatus improved upon Fizeau's, but unfortunately Foucault could not keep the beams of light focused with sufficient accuracy unless the mirrors were fairly close together. This resulted in a small angular displacement of the returning beam. As a result, the relative error of this measurement was rather large—and the first American to win a Nobel Prize did so by adopting Foucault's basic configuration, but devising a way to improve both the absolute and relative errors involved in the measurements.

Albert Michelson—The Man Who Measured Light

Painters throughout the ages have used light in striking and imaginative ways: Tintoretto, de la Tour, O'Keefe (who moved to New Mexico to be able to work its unique light into her paintings), Kinkade (the current self-proclaimed "Painter of Light"). But no one has a greater claim to the use of light as a medium than Albert Michelson, the first American to win a Nobel Prize—which was awarded, as a nominating letter from Professor William Pickering of Harvard stated, "in view of your great work in determining the Velocity of Light and your varied applications of the interference of light."[7]

Michelson's path to the Nobel Prize, and to his study of light, was somewhat circuitous. He was born in the small Polish town of Strzelno. Michelson's family emigrated to the United States, reaching Virginia

City, Nevada, by way of New York and California. Michelson, even as a youth, was a star, and he won a nomination to the U.S. Naval Academy in Annapolis, Maryland.

Whether he would win an appointment was another story. It was a different era. Michelson traveled to Washington for the final interview, where he presented himself to the interviewer—none other than Ulysses S. Grant, then president of the United States. Grant listened intently as Michelson stated his case, but ended by regretfully informing Michelson that only ten at-large positions were available, and they had all been filled. Grant advised an obviously disappointed Michelson to go to Annapolis and wait to see if one of the ten at-large candidates might prove unable to accept their own appointment. Michelson waited three days, to no avail. Discouraged and almost broke, he boarded a train to return to Nevada—and as he did, he heard a man call his name. It was a messenger from the White House. Grant had been so impressed by Michelson that, at the eleventh hour, he had decided to create an eleventh at-large appointment for him.

It was quite a break, but Michelson's career at the Naval Academy was still checkered. He ranked at or near the top of his class in theoretical subjects such as optics and thermodynamics, but near the bottom in seamanship—which one would suspect would be an essential attribute for a career naval officer. Happily, after serving as a midshipman, he avoided the sea, becoming an instructor in physics and chemistry at the Naval Academy.

It was a senior faculty member at the academy, William Sampson, who was to make Michelson preoccupied with light. Michelson was scheduled to teach an advanced course in physics, and Sampson suggested he begin the course with a new pedagogical technique: the lecture demonstration. Sampson thought that Foucault's revolving mirror determination of the speed of light would make an excellent demonstration, and this idea resonated with Michelson, who had already encountered this experiment; indeed, it had been the subject of a question on his final physics exam.

During the demonstration, Michelson recognized that Foucault's experiment had a flaw (one that had also been pointed out by the French physicist Alfred Cornu).[8] The separation distance between Foucault's mirrors was so small that the returning beam was displaced by less than a millimeter, so errors in measurement would have a disproportionately large effect on the computed velocity. Michelson realized that lengthening the distance between the mirrors would significantly improve the precision of the measurement, as would replacing one of Foucault's mirrors with a flat-plane mirror. Michelson was so taken up with the beauty of the experiment and the possibility of effecting a significant improvement in the final result that he chipped in ten dollars to purchase the mirror. Michelson conducted the experiment ten times, took the average of the result, and concluded that the speed of light was 186,508 miles per second.

Further improvements were not long in coming. Several years earlier Michelson had married Margaret Heminway, daughter of a wealthy New York lawyer. Her father was persuaded to donate $2,000 to the Naval Academy for equipment that enabled Michelson to refine his measurements, and several years later, he arrived at the figure 186,355 ± 31 miles per second.

Michelson's preoccupation with light would last a lifetime. Advances in technology, including an interferometer that he invented for the purpose of improving the measurements of very small distances, enabled him not only to continually improve his results, but also led to an experiment that would have a profound effect on the development of physics.

The Michelson-Morley Experiment

According to the prevailing theory of Michelson's era, the Universe was permeated with an invisible weightless substance with the poetic appellation of "luminiferous ether"; disturbances of this substance resulted in waves of light. The waves were real—they had been conclusively demonstrated by the British scientist Thomas Young in his double-slit experiment.[9] In short, it was believed that the ether was to light as air was to sound—you had to have the former for the latter to be able to propagate. This hypothesis led to a critical prediction: If the luminiferous ether existed, the movement of the Earth in its orbit around the Sun should result in different velocities for light beams traveling in different directions, much as a swimmer can travel fastest if he swims with the current (this example was used by Michelson in explaining the idea to his children). It was to measure this difference in speed that Michelson and Edward Morley, a professor at what is now Case Western Reserve University in Cleveland, constructed an exquisitely beautiful and conceptually simple experiment.

The Michelson-Morley experiment consisted of splitting a beam of light in two perpendicular directions to two different mirrors located at the same distance from the point at which the beams diverged. The light waves would come back—and, assuming the ether was real, at different velocities—and the waves would interfere with each other. Michelson's device, known as an interferometer, could be used to determine the difference in speeds between the two returning waves, which would enable them to compute the speed with which the Earth was traveling through space. So sensitive was the interferometer that someone stamping their foot 100 feet away would register, rendering whatever results they obtained invalid. The interferometer and the beam splitter were placed on a slab of marble that floated in a pool of mercury—one can envision Indiana Jones tiptoeing up to this in a dark cave to snatch a priceless relic. This arrangement helped shield the equipment from disturbance, and had the added advantage that the slab could be rotated on the mercury pool to produce results at many different orientations. According to Eddington, the device could measure a difference of one ten-thousandth of a billionth of a second in the return times of the light beams—a time during which light travels a little more than one one-thousandth of an inch.

The result they obtained astounded the physics community—no matter how often they repeated the experiment, the waves returned at exactly the same time. The conclusion was difficult to accept—the speed of light was the same in any direction. It was rather like learning that a swimmer's rate is the same no matter whether he swims with the current or against it.

There were several possible conclusions. The most commonly cited one is that the failure to detect a difference in the speed of light no matter which way the beams were aligned showed that the luminiferous ether could not exist; if it did, there would have been an alignment of the beams resulting in a detectable variation of the speed of the returning beams. However, the Irish physicist George FitzGerald came up with a surprising explanation of the "null result" of the Michelson-Morley experiment. He hit upon the seemingly bizarre hypothesis that when an object moved through space, its length shrunk in the direction that it moved by just enough to ensure that the paths of both beams of light returned at the same instant. This phenomenon, the FitzGerald contraction, was amusingly described in the following limerick.

> *There once was a fencer named Fisk,*
> *Whose fencing was startlingly brisk,*
> *So fast was his action,*
> *The FitzGerald contraction,*
> *Reduced his épée to a disk.*

The Dutch physicist Hendrik Lorentz was able to quantify this phenomenon algebraically in equations known as the Lorentz transformations. In his special theory of relativity, Albert Einstein was able to derive the Lorentz transformations under the twin assumptions that the speed of light was constant in all reference frames that moved at a constant speed (another possible conclusion of the Michelson-Morley experiment) and the relativity assumption that the laws of physics are the same in all such reference frames.

Although Michelson's work centered around light, there were dark episodes in his personal life that provide a counterpoint. Michelson lived in an era in which professors, especially well-known ones, were minor celebrities—a phenomenon that continued into the early part of the twentieth century. Michelson hired an attractive but relatively simple-minded maid who allowed herself to be used in a plot to extort money from Michelson by charging him with attempted seduction (you don't see this on police blotters much anymore), assault, and battery. A scandal ensued, but Michelson was exonerated. There was worse, however. Michelson, like many geniuses in diverse areas of human endeavor, was a workaholic, which led to a nervous breakdown and eventually to the dissolution of his marriage. Although he later remarried, his divorce embittered him, and he never spoke to or of his first wife and children after that experience. His students admired his brilliance but feared his intractability, a feeling shared by his colleagues. He enjoyed painting and composed music, yet neither of these activities produced much softening in the severe exterior he presented the world. His research assistant of many years summed him up by writing, "Even those cosmic human forces of love, hate, jealousy, envy, and ambition seemed to move him little. He possessed an astonishing indifference to people in general. . . ."[10] Nonetheless, he was passionate in his pursuit of the nature and attributes of light. His daughter mentioned that when someone asked him why he had spent his entire life in the study of light, his face lit up as he replied, "Because it's so much fun."[11]

Faster Than a Speeding Photon

Part of what makes light so much fun is not just that it is a deep constant of the universe—there's also the fact that it is an upper limit on action in the universe. Nothing can travel faster than light—not even information—and nothing with any mass can even go as fast as light.

Well, let me complicate that claim a bit. Picture a lighthouse located on some rocks some distance out to sea from a beach. The beach has a sea wall behind it, and as the light in the lighthouse turns, the light shines upon the wall. The light on the wall appears to move: fairly slowly as the beam is pointed perpendicular to the wall, more rapidly as the beam rotates toward a direction parallel to the wall. Here's the fun part—despite everything I wrote in the previous paragraph, we can show fairly easily that under the right conditions, the speed with which the light beam moves down the wall can exceed the speed of light itself!

Making this a geometry problem will show us how. If you're not familiar with it (or even if you are), bear in mind the Pythagorean Theorem, which states that for any right triangle with sides a and b and hypotenuse c, the sum of the squares of the sides a and b equals the square of c (so $a^2 + b^2 = c^2$). Now, let's assume the lighthouse is located at L, which is a distance R from the wall. The point Y will represent the point on the wall that is at distance R from the lighthouse. At some future time, the beam of light will have moved to a point X on the wall, which we will assume is at distance d from Y. The points X, Y, and L form a right triangle whose legs are of length R and d.

Assume that at time $t = 0$ the lighthouse emits a light beam directly at Y; because time = distance / speed, the light beam will hit Y at $t = R / c$ (where c, as is common practice, denotes the speed of light). Also assume that it takes q seconds for the lighthouse to rotate one quarter of a revolution. It will take less than q seconds for the lighthouse to rotate to a position where the beam that it emits will hit X, and when it rotates to that position, the beam will travel the hypotenuse LX of the triangle, a distance of $\sqrt{R^2 + d^2}$. It will take $\sqrt{R^2 + d^2}\ / c$ seconds for the beam to travel from the lighthouse to X. Consequently, the beam will hit X before $t = q + \sqrt{R^2 + d^2}\ / c$ seconds, and so the beam will have moved the distance d from Y to X in less than $s = q + \sqrt{R^2 + d^2}\ / c - R / c$ seconds. In order for the beam to cover this distance in less time than it would take light to cover the same distance, we must have $d / s > c$. Multiplying both sides of this inequality by s, we see that we would need

$$d > (q + \sqrt{R^2 + d^2}\ / c - R / c)\, c = qc + \sqrt{R^2 + d^2} - R.$$

A little addition and subtraction gives the following form for the needed inequality as

$$R - qc > \sqrt{R^2 + d^2} - d$$

Now comes another algebraic ploy; we multiply both sides by $\sqrt{R^2 + d^2} + d$; we've basically multiplied an expression of the form $a - b$ by $a + b$ to yield $a^2 - b^2$. In our case, the needed inequality now becomes:

$$(R - qc)\, (\sqrt{R^2 + d^2} + d) > (R^2 + d^2) - d^2 = R^2.$$

To make this happen, we first make sure that the lighthouse is rotating so rapidly that the expression on the left-hand side is positive (this can be done by choosing a rotation rate for which $q < R / c$). Once this is done, if we can choose d such that $(R - qc) (2d) > R^2$ we'll have succeeded in our quest, because $\sqrt{R^2 + d^2}$ is always greater than d (the hypotenuse of a right triangle is always longer than each of the legs), and so $\sqrt{R^2 + d^2} + d > 2d$. Therefore, as long as we first choose a rotation rate such that $q < R / c$ and then choose a value for d for which

$$d > R^2 / (2(R - qc))$$

we'll have constructed a situation in which the beam of light moves along the wall faster than the speed of light itself!

Let's get an idea of the speeds and dimensions involved by looking at a real example. An automobile shaft typically can rotate at a speed of 6,000 rpm if the pedal is pushed to the metal, this would be a speed of 100 revolutions per second, and so a quarter of a revolution would take 1/400 of a second. That's the value for q. Since the speed of light is about 186,000 miles per second, $qc = 186,000 / 400 = 465$ miles. We need $R > qc$, so let's take R to be 500 miles (okay, so the lighthouse is a long way from the beach). We then need $d > 500^2 / 2 (500 - 465)$, or greater than roughly 3,570 miles (okay, so it's a long beach—even longer than those near my home institution of Cal State Long Beach).

If you're interested in refining this calculation so the numbers appear to be more down-to-earth, there are at least two ways to go about it, both of which rely on the fact that the above computation results in an *average* velocity faster than the speed of light over the entire length of *XY*. We know from experience that the beam of light appears to move faster the closer it comes to completing the quarter revolution that will make the beam parallel to the shore, so one could either compute the average velocity over an interval *ZX*, where the point *Z* is very close to *X* (this doesn't require trigonometry, but trig makes it easier), or compute the instantaneous velocity of the moving point at *X* (this requires calculus).

Don't think that this violates the principle that nothing can move faster than light. The moving point of light is neither a thing nor an electromagnetic wave—it's the intersection of a light beam with a wall, which is essentially a mathematical construct. If instead of using a light beam, we had a rotary paint-sprayer, what would be moving down the wall is the end of the line of paint that started at X; no individual paint particle would be moving at that speed. For those who prefer to keep light in the picture, simply having the wall consist of a detector that permanently records the impacts of photons would produce a line similar to the line of paint. Finally, the quantity c represents the speed of light in the calculation; it appears on both the left and right sides of the inequality $d / (q + \sqrt{R^2 + d^2} / c - R / c) > c$. On the left side of the inequality it represents the speed at which the light beam travels from the lighthouse to points on shore, but on the right side of the inequality it represents the speed of light along the beach: the velocity we are trying to exceed. If instead of c on the left-hand side of the inequality we were to use p, the speed of paint, we would obtain the inequality $d / (q + \sqrt{R^2 + d^2} / p - R / p) > c$. This inequality can be satisfied for much lower values of p than c; but to do so we need to position the lighthouse much further from the beach and we need to elongate the beach considerably. Pretty neat—nothing can go faster than the speed of light, but if you look for the correct nothing, well, there it is!

CHAPTER 3

THE
IDEAL GAS
CONSTANT

Science today consists of observation, experimentation, and theorizing, but it hasn't always. In ancient Greece, for example, there was little or no means for experimentation, and so the scientists of the time, who called themselves philosophers, were confined to observation and theorizing in order to answer the question of how the universe and the things in it came to be. One of the earliest was Thales of Miletus who made the first successful prediction of an eclipse of the Sun and also may have produced the first proof in geometry when he showed that vertical angles were equal. Thales theorized that all things came from and depended upon water. A century or so worth of elaborated speculation culminated with Empedocles's declaration that all things were in fact produced by mixing and separating the four elements of earth, air, fire, and water; the mixing and separating taking place under the influence of two forces, which Empedocles described as "love" and "strife."

It seems primitive from the standpoint of the twenty-first century, but they didn't do so badly! There are four phases of matter—solid (corresponding to earth), liquid (water), gas (air), and plasma (fire). I wouldn't want to stretch this too far, but the electromagnetic force on which chemistry depends could be viewed as "love" (opposite electric charges or magnetic poles attract) and "strife" (like electric charges or magnetic poles repel). Throw in the atomic theory originally elucidated by Leucippus and Democritus more than four centuries before Christ, and you don't have too bad a job of describing the universe and the things in it. The progress of science has thrown out a lot of the bad guesses and kept the good ones, which were pretty good, considering that they didn't have much to work with.

Two millennia later, in seventeenth-century Europe, the philosophers' heirs—then known as natural philosophers—hadn't abandoned observation and theorizing, but they had developed some means for experimentation, and were going at it hammer and tongs, which happen to be two of the implements needed for constructing the means for experimentation. It was time to get down and dirty with nature.

Most scientists, I think, would agree with this definition of *experiment*: the act of conducting a controlled test or investigation. When this sort of thing began is open to some dispute. Galileo had used experiments to conclude that the distance an object fell under the influence of gravity was proportional to the square of the time the object was falling, and had done so by rolling balls down inclined planes, in the early seventeenth century. A few decades later, Anton von Leeuwenhoek was putting practically everything under the microscope—not metaphorically but actually. But it was Robert Boyle, a contemporary of von Leeuwenhoek's, who developed the methodology we think of as the quintessence of experimentation: varying one parameter and seeing how other parameters change in response. He kept a journal in which he recorded the apparatus used, the procedures involved, and the measurements observed, thereby laying the foundation for experimental science.

A physical law is often codified in a mathematical relationship (usually an equation, occasionally an inequality) that describes how the changes of the various parameters involved are related. In order to obtain such a law, it is necessary to have changes to observe. Of the four phases of matter, gases are the easiest in which to observe and measure such changes; liquids and solids don't change that much (at least, given the sensitivity of seventeenth-century measuring equipment), and plasma as a form of matter was unknown in that era. So perhaps it was a foregone conclusion that the first laws to be derived concerning matter would involve gases. At any rate, that's where Boyle began.

He was not alone in the study. In 1643, the Italian physicist Evangelista Torricelli had discovered that a column of air sufficed to support a column of mercury that we would describe as 760 millimeters high, and that the atmospheric pressure could change. This inspired the German scientist Otto von Guericke, who was also the mayor of the town of Magdeburg, to create the first vacuum pump. To demonstrate the power of atmospheric pressure, he devised what are now known as the Magdeburg hemispheres: two copper hemispheres approximately 20 inches in diameter that incorporated seals so the hemispheres could be evacuated by means of the pump. When this was done, a team of horses was unable to pull the hemispheres apart; the hemispheres fell apart by themselves when air was allowed to reenter the hemispheres.

It's easy to compute how much force would have been needed to separate the hemispheres. A column of air whose cross section is 1 square inch weighs approximately 14.7 pounds, and the surface area of a sphere of radius R is $4\pi R^2$. The total force required to separate the two hemispheres, given a radius of 10 inches, is therefore $4 \times 3.14 \times 10^2 \times 14.7 = 18,463$ pounds. This, in turn, is equal to the force of the atmosphere upon the joined hemispheres.

You may be a little skeptical about the above calculation, yet there is a simple experiment you can perform that should convince you that the total force is the atmospheric pressure times the surface area. Fill a glass with water, and find a plastic lid (paper won't do, because it gets soggy) that barely covers the glass. Make sure the lid is completely dry, and cover the glass, then turn it upside down. The atmospheric pressure is greater than the weight of the water, and the lid stays on! *Do* try this at home!

When news of the von Guericke demonstration reached Boyle, he resolved to build a simpler vacuum pump. Von Guericke's pump had required two men to operate it; Boyle's improved version could be easily operated by just one individual. Boyle's investigations into the nature of air were first published in "The Spring and Weight of the Air."[1]

In the first edition, published in 1660, Boyle showed that sound could not be transmitted in a vacuum by trying to ring a bell in an increasingly rarefied chamber, and that air was necessary for both life and the maintenance of a candle flame. It was the second edition, published in 1662, that contained the relationship between pressure and volume that every introductory physics or chemistry student learns as Boyle's law. Boyle's law states that as long as the temperature is kept constant, pressure and volume vary inversely with each other—there is a constant k such that $PV = k$, where P is the pressure in the gas and V its volume.

Boyle was fortunate to have Robert Hooke as his lab assistant. Hooke was the first of many lab assistants who would go on to do significant scientific work on their own. Working in Boyle's laboratory enabled Hooke to pick up on Boyle's methodology, which he employed in establishing Hooke's law (the restoring force on a spring is proportional to the length that the spring is stretched beyond its natural length). Possibly he was influenced to undertake these investigations because of Boyle's belief in the springiness of the air. Hooke is also responsible for one of the most important observations in the history of science. In 1665, while examining thin slices of cork under the microscope, he discovered that they were made of cells (the term "cell" comes from Hooke's comparison between the compartments visible in the cork slice and the small rooms—cells—occupied by monks in a monastery).[2]

There is some disagreement among historians as to the role Hooke played in the discovery of Boyle's law. Hooke appears to have conducted some of the experiments, and one historian[3] suggests that Hooke, who (unlike Boyle) was an accomplished mathematician, may have developed the mathematics of Boyle's law. At any rate, Boyle and Hooke had considerable respect for each other, and Boyle was not a person to stand in Hooke's path. When the newly founded Royal Society needed a curator for its experiments, Hooke was well known to all its members, and was unanimously awarded the position.

Regardless of how the credit for Boyle's law is apportioned, Boyle was unquestionably one of the leading scientists of his day. His influence extends beyond the physics of Boyle's law, the discoveries he made in his investigations of air, and his considerable contributions to chemistry. It was Boyle who first established the basic approach that characterizes empirical science, and one of the individuals to adopt his approach was Isaac Newton.

John Dalton

Even though the physical state of a gas can be described by only three parameters—pressure, volume, and temperature—it took more than a century after Boyle's law was discovered for a relationship between temperature and volume to be determined. There's a very simple reason for this—during the seventeenth century, there was no method to measure temperature. Once such a method was devised, several prominent scientists tackled the relationship between these quantities.

One was John Dalton, one of the most important figures in the history of science. Dalton is generally credited with the development of the atomic theory—that the primary constituents of matter are atoms, and that chemical compounds are formed by the combining of atoms of one element with atoms of another element. Dalton also investigated the relationship between the temperature and the volume of a gas, and reached the conclusion that if the pressure was kept constant, the gas expanded at a constant proportion of its volume at the preceding temperature.[4] Generally a meticulous experimenter, Dalton was somewhat sloppy in this particular effort—and as a result reached the wrong conclusion. Although Dalton was correct that the expansion of a gas was a constant proportion of volume, his conclusion that this constant proportion was of the *preceding* volume was erroneous.

To understand Dalton's conclusion, let's assume that Dalton's proportionality constant was 1 percent per degree centigrade—for every degree centigrade that a gas was heated at constant pressure, it would expand 1 percent of the preceding volume. Suppose that at 0 degrees centigrade, the gas occupied a volume of 10,000 cubic centimeters. Using 1 percent per degree centigrade as the coefficient of expansion, if the gas were heated to 1 degree centigrade at constant pressure, it would expand 1 percent to 10,100 cubic centimeters. If the gas were heated another degree at the same pressure, to 2 degrees centigrade, it would expand 1 percent of the preceding volume of 10,100—to a new volume of 10,201 cubic centimeters. If gas were money, Dalton's conclusion was that thermal expansion acted the same way as compound interest does—that extra 1 cubic centimeter, the difference between 1 percent of 10,100 and 1 percent of 10,000, is the "interest on the interest" that is the hallmark of compound interest.

However, there's a problem here that I haven't seen specifically mentioned in the literature. I'm not enough of a history wonk to really do a thorough job of researching this, so it's certainly possible that someone—maybe even Dalton—recognized the difficulty. Compound interest depends on the number of compounding periods in a given year. There is a formula that is learned by most intermediate algebra students and—I hope—all business majors. This formula is $A = P(1 + r / N)^{Nt}$, where P is the amount deposited, r is the annual interest rate expressed as a decimal, N is the number of compounding periods per year, and A is the amount in the account after t years.

If $10,000 is invested at 1 percent compounded annually, the amount in the account at the end of one year is $1.01 \times \$10,000 = \$10,100$. However, if the 1 percent is compounded semiannually, the 1 percent is split between the two half-year compounding periods. Using semiannual compounding, the amount in the account at the end of six months is $1.005 \times \$10,000 = \$10,050$, and the amount in the account at the end of one year is $1.005 \times \$10,050 = \$10,100.25$; we could also have used the formula given previously with $P = 1,000$, $r = .01$, $N = 2$, and $t = 1$. If we accept Dalton's conclusion that the amount of gas increases in proportion to the volume at the preceding temperature, we run into a problem stemming from the two different ways of doing the computation. Is the volume of the gas at 1 degree centigrade computed as a percentage of the gas at 0 degrees centigrade (analogous to annual compounding), or do we do it in two stages, by first computing the volume of the gas at 0.5 degrees centigrade as a percentage of the volume of the gas at 0 degrees centigrade, and then computing the volume of the gas at 1 degree centigrade as a percentage of the volume of the gas at 0.5 degrees centigrade (analogous to semi-annual compounding)?

Even worse, it's possible to compound quarterly, monthly, or daily—and the balance at the end of the year is different for each of these methods of compounding. There's a way out—but again, I've seen no reference to this in my admittedly meager historical searches. If one goes through the process of compounding through ever shorter periods: every month, every day, every hour, every second, every nanosecond . . . the amount in the account at the end of the year gets larger, but it increases to a limiting value. It increases exponentially according to what is referred to as the PERT formula. If a principal P is deposited in a bank for t years at a decimal rate r compounded continuously (the result of increasing the frequency of the compounding periods beyond nanoseconds, picoseconds, what have you), the amount A in the account is given by $A = Pe^{rt}$, where e is the base of the natural logarithms. The corresponding PERT formula (an acronym deriving from the right-hand side of the previous equation) for gases would yield a final volume of A for a given initial volume of P, a temperature increase of t, and a yet-to-be-determined constant of nature r that only experimentation and measurement will reveal.

The final amount in a bank account subject to continuous compounding depends only on the amount deposited, the interest rate, and the length of time that the money is in the account. The final amount in a bank account subject to periodic compounding depends on all these factors, but also on how frequently we compound. The analogy for gas expansion would be that if gas expanded at constant pressure according to a PERT formula, the volume of gas present at the end of the expansion would depend only on the initial volume, an expansion rate (comparable to the interest rate) that is a constant of nature, and the initial and final temperatures. If gas expanded at constant pressure according to some sort of periodic rate, the final volume would depend upon all these factors *and* the procedure by which the gas was heated to its final temperature—whether the temperature was raised continuously, or 1 degree at a time, or something else. This difficulty would be the logical result of a Dalton law of expansion. Although this difficulty would be avoided if the law of expansion adhered to a PERT formula, such an adherence was not what experiments showed.

The PERT formula that one sees in continuously compounded interest is a specific example of what is known as exponential growth and decay—and this does occur quite commonly in nature. Exponential growth (or decay) occurs when the rate of growth (or decay) is proportional to the population. I started teaching back in the 1960s when *Star Trek*—the version with Kirk and Spock—was exceptionally popular with college students. One of the most amusing episodes was called "The Trouble with Tribbles"; tribbles were attractive furry creatures that reproduced extremely rapidly, and soon the *Enterprise* was overrun with tribbles. As Bones McCoy, the ship's doctor said, they reproduced so quickly that tribbles were probably born pregnant. If you have twice as many tribbles to start with, you have twice as many baby tribbles— the rate of growth of the population is proportional to the population. I used tribbles to illustrate exponential growth problems; I even brought a couple of dormant tribbles to class. (Actually they were powder puffs purchased from a nearby drugstore, but *Star Trek* was produced in an era with nothing like computer-generated special effects, and the tribbles on the show looked remarkably like powder puffs.) Radioactive decay operates similarly; if you have twice as much of a radioactive substance to start with, twice as much radioactive material will turn to lead. You may think that the phrase "knowledge increases exponentially" is a figure of speech, but it's actually a literal description of what happens if the rate of acquisition of knowledge is proportional to the amount of knowledge in existence.

There was an indication, though, that Dalton's conclusion was in error. Just as more interest accrues to an account subject to compound interest after the interest from early compounding periods is added to the account, Dalton's conclusion would result in gases undergoing more expansion at higher temperatures than at lower ones. Experimentation did not seem to bear this out, and the problem was eventually resolved by two French pioneers in the budding eighteenth-century aerospace industry.

The French Balloonists

The correct formulation of the law governing gas expansion while the pressure remained constant was reached by two Frenchmen: Jacques

Alexandre César Charles and Joseph Louis Gay-Lussac. Both were scientists, but each was also interested in the expansion of gases because of their interest in a cutting-edge technology of the late eighteenth century: hot-air ballooning. Heat made gases expand, and too much expansion of the gas in a hot-air balloon could result in the balloon rupturing, with the obvious catastrophic results. Knowing how much heated air would expand would obviously help avoid such occurrences. Gay-Lussac was interested not just in ballooning for its own sake, but in investigating the nature of the atmosphere at higher elevations. In 1804 he and a fellow scientist reached an altitude of about 23,000[5] feet in order to obtain temperature and moisture measurements at different elevations—possibly a world altitude record for that era. Although there was no *Guinness Book of Records* at the time to validate it as such, it does not appear to have been surpassed for almost fifty years.

What both Charles and Gay-Lussac discovered was that nature did not compound thermal expansion; it instead charged simple interest as a gas was heated at a constant pressure. Their conclusion can be seen in contrast to Dalton's. The principal in both cases is the volume of gas at 0 degrees centigrade. If the proportionality constant was 1 percent per degree centigrade, both theories predicted the expansion of the gas to 10,100 cubic centimeters as the gas was heated from 0 degrees centigrade to 1 degree centigrade. However Charles and Gay-Lussac stated that when the gas was heated from 1 degree centigrade to 2 degrees centigrade, the gas would expand by an additional 1 percent of volume at 0 degrees centigrade—from 10,100 cubic centimeters to 10,200 cubic centimeters.

Well, not quite: I have used 1 percent per degree centigrade as an example because it makes for easier calculation, but the number deduced by Gay-Lussac was that gas expanded by 1/266.67 of its volume at 0 degrees centigrade per degree centigrade of heating. Gay-Lussac's coefficient of expansion was very close to the currently accepted coefficient of 1/273.15; which is a tribute to his ability as an experimenter.

This result is sometimes known as Gay-Lussac's law, but more commonly as Charles's law—and it is known as Charles's law thanks to the efforts of Gay-Lussac! Charles obtained his results fifteen years prior to Gay-Lussac, but Gay-Lussac did a more precise job of documenting his results, and did what scientists and mathematicians generally do (but Charles didn't); he published his results.[6] He also credited Charles with having done the experiments. In consequence, the result on the thermal expansion of a gas at constant pressure is known in some quarters as the law of Charles and Gay-Lussac.

Putting It All Together

At the beginning of the nineteenth century, scientists had two laws concerning the behavior of ideal gases. Boyle's law stated that if the temperature were kept constant, the relation between pressure and volume was given by $PV = k$ (for some constant value of k). Charles's law, or Gay-Lussac's law, expressed a similar type of relationship between the absolute temperature and volume of a gas that was being kept at constant pressure—this relationship was given by $V = Tk'$, where k' was also a constant—but a different constant from the constant k that appeared in Charles's law.

It wasn't until the middle 1830s that the physicist Émile Clapeyron fused these two laws into what we now know as the ideal gas law, which I find surprising, because the relationship between the two seems fairly straightforward. When we consider Boyle's law, the constant k that appears on the right-hand side of the equation will be different for different absolute temperatures T, so we can express it as a function of T, which we write as $f(T)$. Boyle's law is now rewritten as $PV = f(T)$.

If we apply the same reasoning to Charles's law, the constant k' that appears on the right-hand side of the equation will be different for different pressures P, so we can express it as a function of P, which we write as $g(P)$. Charles's law now becomes $V = g(P)T$.

It's time for a (very) little algebra: $f(T) = PV = Pg(P)T$. If we divide both sides of the equation $f(T) = Pg(P)T$ by T, we obtain the relation $f(T) / T = Pg(P)$ for all values of T and P. However, the expression on the left, $f(T) / T$, depends *only* on the value of T, whereas the expression on the right, $Pg(P)$, depends *only* on the value of P. Imagine now that we heat the gas up but keep its pressure constant, say $P = P_0$. If the expression $f(T) / T$ changes, then $P_0 g(P_0)$ would have different values, which is clearly impossible. Consequently, $f(T) / T$ must have a constant value, which we will abbreviate by the letter a. Since $f(T) / T = a$, $f(T) = aT$, and therefore Boyle's law becomes $PV = f(T) = aT$.[7]

A little thought shows that the constant on the right really isn't a constant; it depends on how much gas we had at the start of the experiment. Suppose that we were to build a large container and partition it into two equal parts with a removable divider. Conduct the same experiment on each side of the divider—the pressure P, the temperature T, and the volume V will be the same on each side. Remove the divider, and the pressure P and temperature T do not change—but the volume V doubles. So $PV = aT$ and $P(2V) = 2PV = 2aT$, which shows that the constant on the right side of the equation doubles as the volume doubles. Similarly, if we had a container divided into three equal parts, we would see that the constant on the right side of the equation triples as the volume triples. The constant on the right-hand side therefore depends on how much gas we start with. This is incorporated in the final form of the ideal gas law, which is written $PV = nRT$. The constant n denotes the amount of gas present, which is usually measured in moles (this quantity will be described in the chapter on Avogadro's number, for those who either haven't seen it or have forgotten their high school chemistry), and R is the ideal gas constant.

Statistical Mechanics and the Ideal Gas Law

Physics, like the other sciences, seeks not only truths but explanations. Although the ideal gas law is derived from Boyle's and Charles's laws, both of these laws are empirical, the result of experimenting, observing, and measuring. The goal, as with Newton's investigations of Kepler's laws, was not just to have an empirical description of nature, but a theoretical framework that explained it—to enable us to accurately predict what we would observe empirically from purely theoretical underpinnings. The latter half of the nineteenth century witnessed such a great leap forward in the understanding of the behavior of gases with the advent of statistical mechanics.

Statistical mechanics is the application of probability theory (a branch of mathematics that includes statistics) to thermodynamics; it considers an ideal gas to be a collection of a very large number of particles (the atoms or molecules of the gas) with positions and velocities that are given as probability distributions. It isn't always possible to know where a specific gas molecule is, but one can assume that at any given moment, its position in a glass jar is random—it's just as likely to be near the top as near the bottom. It is possible to derive the ideal gas law from the assumptions of statistical mechanics using some fairly high-powered tools from both mathematics (the divergence theorem, one of the crown jewels from multivariable calculus) and physics (Newton's laws of mechanics, Hamilton's equations, and the equipartition theorem[8]). This puts the ideal gas law on the same firm theoretical foundation (if such a foundation can be described as *firm*) as Kepler's first law.

From a practical standpoint, more complete scientific theories are advantageous for a number of reasons, not the least being that they suggest new technology. They also allow scientists to deduce how things will behave without having to perform the experiments, except to confirm the predictions. From an epistemological standpoint (this is the first time in nearly seven decades that I've used the word and I probably never will again), a more complete scientific theory pushes back the veil of ignorance—but may never completely remove it. Showing that Kepler's first law is a mathematical consequence of Newton's theory of gravitation replaces the question "Why do planets travel in ellipses with the Sun at a focus?" with the question "Why does the gravitational force act on a straight line between two bodies at a strength proportional to the product of their masses and inversely proportional to the square of the distance between them?" Einstein hoped to remove this veil of ignorance by producing a unified field theory. Modern physicists hope to accomplish this with a TOE—a theory of everything.

I'm skeptical. One of the advantages of writing a book such as this is that you get to interject your own opinions. I believe that for an infinitesimal instant after the big bang all the four forces were unified—and that's the last time there was a simple underlying explanation for anything. Many phenomena fall into explainable patterns—but I do not necessarily believe that these explainable patterns form patterns of their own, regressing to one simple pattern that explains everything. I concur with Richard Feynman in believing it would be incredibly lovely if such were the case, but I think the universe in its totality is like Feynman's onion, with layers that are continually peeled back to reveal deeper truths. Maybe physics has some unity, but that's only because physics deals with phenomena that are the essence of simplicity when compared to complex chemistry, cellular biology, or the human brain. Bring in any level of complexity, and hardly anything is predictable— or simple.

This won't dissuade physicists—or others, for that matter—from looking for such an explanation; they do so for the same reason that Albert Michelson spent so much time and effort measuring the speed of light—because it's so much fun.

CHAPTER 4

ABSOLUTE
ZERO

_____

W hen I was growing up, Sunday afternoons were often devoted
to what nowadays might be called intellectual enrichment. My
suburban parents took advantage of the abundance of museums that
could be found in New York, and after a short train trip on the New
York, New Haven, and Hartford we would have lunch at Schrafft's and
head for a museum. I suffered through trips to the Metropolitan Museum (except for the suits of armor, because it was amazing how _short_
knights of the fifteenth century were), but I would put up with them
because I knew that next time my parents would take me to the American Museum of Natural History or, if I was really lucky, the Hayden
Planetarium. Even today—despite the wonders of the Internet—it's
hard to believe that web surfing could produce the same sense of wonder in a child as a trip to the planetarium.

Every moment at the Hayden Planetarium was to be savored. It
would end with the Zeiss planetarium projector show, which was better
than anything I ever saw at the movies, but another treat was walking
through the section of the Planetarium devoted to the solar system. You
could find out how much you weighed on Mars or what the temperature
was on Pluto (for me, Pluto will _always_ be a planet). The big question
for me was—how do they know?

I knew we hadn't been to Pluto, or even to Mars—and how can you measure the temperature of a place you haven't been? And even if you could go to Mars or Pluto, how could you ever go to the surface of the Sun, where the temperature was said to be 6,000 degrees? The blue giant stars were even hotter, with surface temperatures of 50,000 degrees. But these were dwarfed by the temperature of the solar corona, about 1,000,000 degrees, and the core of the sun had temperatures of 25,000,000 degrees. How did they know? What did a thermometer that registered 25,000,000 degrees look like?

There was an even odder thing I wondered about: 25,000,000 degrees was a long way from the temperatures in suburban New York, but 400 degrees below zero, roughly the surface temperature of Pluto (once a planet, always a planet), didn't seem so far. Why did temperatures get incredibly hot, but didn't seem to get much colder than 400 degrees below zero?

The Nature of Cold

The question of whether there is an ultimate limit to cold seems to have first put in an appearance in the seventeenth century. The seventeenth century featured two similar theories on the cause of heat and cold. The phlogiston theory of heat viewed flammability in terms of possession of a substance called phlogiston; when a material was burned, the air absorbed the phlogiston and the burnt substance became "dephlogisticated." Cold was likewise transferred from one substance to another; the frigorific theory (which sounds like a 1930s trade name for a refrigerator) held that there was an ultimate cold body called the *primum frigidum*.[1] This body was the ultimate dispenser of cold; all other bodies acquired their coldness from it.

The talented British physicist Robert Boyle was one of the first to investigate the nature of cold through scientific experimentation. It was probably not a coincidence that this occurred during the Little Ice Age, when Europe was undergoing a multi-century cold snap that reached a minimum during the period of Boyle's investigations. Boyle weighed a barrel of water, left it out in the snow, and then weighed it the next day when the water had frozen. Ice occupies a larger volume than the water needed to create it (as can be seen from the ice cubes floating in a glass of water), and even though the expanding ice had broken the barrel, the ice weighed the same as the water. If water was absorbing something from the *primum frigidum*, what it absorbed clearly didn't weigh anything (at least to the limit of seventeenth-century weighing devices). Boyle reached the conclusion that substances became hot and cold because of some internal characteristic of the substances. However, further developments in thermodynamics had to wait upon the development of thermometry and the adoption of a calibration scale.

The Development of Thermometry and the Experiments of Guillaume Amontons

Hero of Alexandria, the inventor of the steam engine (although not one that could be commercialized, or the Industrial Revolution might have occurred almost two millennia earlier), was aware that air expanded when heated, and created a primitive thermometer by immersing the open end of a tube of air in a container of water. As the air expanded or contracted, the air-water boundary moved. The problem was that this apparatus was also sensitive to changes in air pressure. A way to circumvent this was found in the middle of the seventeenth century, when Fernando II de' Medici, the Grand Duke of Tuscany, used a sealed tube of alcohol instead of air; shielded from atmospheric pressure, the volume of the alcohol depended only on the ambient temperature. The inclusion of a graduated scale next to the tube made the thermometer truly useful, and when alcohol gave way to mercury as the indicator of choice for scientists, the modern thermometer's development was pretty much complete. Mercury enabled more compact thermometers to be constructed. It was with such an air-mercury thermometer that Guillaume Amontons first suggested the possibility of a numerical value for the lowest possible temperature.

Amontons was a French scientist who was deaf from childhood, which probably prevented him from attending a university. He studied mathematics and science on his own, made improvements in the thermometer, barometer, and hygrometer, and turned from the study of friction to an investigation of the relationship in a gas between temperature and pressure. He immersed a container of air joined to a column of mercury in water. Amontons noted that the temperature corresponding to the melting point of ice was 51 on his scale, and the temperature corresponding to the boiling point of water was 73. Amontons argued that when the pressure (and corresponding volume) of the air was zero, no further cooling could take place, although he did not "do the math" required to determine the temperature of absolute zero.

We, however, can do the math that Amontons did not. One unit on Amontons's scale is equal to $8\frac{2}{11}$ Fahrenheit degrees. Given his data, a reduction of 51 units, or 417 Fahrenheit degrees, from the freezing point of water would be required to reduce the pressure to zero; this corresponds to an estimate of −385°F for absolute zero. This isn't so bad (the actual value is about −459.67°F), especially considering that Antoine Lavoisier, Pierre-Simon Laplace, and John Dalton, three brilliant scientists, would conclude in the latter portion of the eighteenth century that absolute zero lay between −1,500°F and −3,000°F.[2] The actual value, as well as the name, of absolute zero would eventually come from Lord Kelvin, who developed a scale starting at absolute zero that placed the freezing point of water at 273.15 degrees—precisely where the modern system of measurements, which records temperatures in kelvins, puts it today.

Despite the theoretical success, no one was considering an assault on absolute zero; at this time it would have seemed as distant as the Moon or the stars. This was undoubtedly due to the fact that there was no way to create cold. Creating heat was easy—find something flammable and burn it. But only nature could create cold, until an experiment conducted by a scientist who was better known for practically everything else he did.

Michael Faraday and the Liquefaction of Gases

Michael Faraday was a young man working for a bookbinder when he heard a series of lectures given to the public by the eminent chemist

Sir Humphrey Davy. Faraday took a shot, and wrote Davy to ask if he needed an assistant. Davy was impressed, and hired Faraday. Thus began the career of one of the greatest experimenters in the history of science.

Faraday is undoubtedly best known for his experiments involving electricity, but he also made significant contributions to chemistry. One day he had produced liquid chlorine from chlorine hydrate. The chlorine was in a sealed tube, and Faraday decided to examine it more closely. He broke the tube—and the tube exploded into shards of glass, which flew around the laboratory. The chlorine was instantly transformed from a liquid into a gas. Faraday, ever the adroit observer, noted that if vaporization of a liquid resulted in an explosion, possibly the reverse—applying pressure to a gas—would produce liquefaction. Interestingly enough, although Faraday was interested in pure science rather than commercial applications of scientific discoveries, he did note that someday commercial applications might be found for this particular phenomenon.[3] That, of course, happened—and the refrigerator in your kitchen is a primary example. A liquid circulates throughout a system of coils in your refrigerator. It evaporates in a chamber inside the refrigerator, which causes heat to flow into the evaporating liquid from the surrounding environment, cooling it. The gas is then pressurized outside the refrigerator by an electric pump; this liquefies it and releases the heat that the liquid absorbed in the refrigerator. The cycle continues until an equilibrium temperature is reached. Simple enough, but the discovery revolutionized the world.

Chemists display the information about where various substances liquefy and solidify through a phase diagram. One axis represents the pressure P, the other axis represents the temperature T, and the P-T plane divides into regions, each of which represents a phase of the substance.[4] The phase diagram for many gases, such as carbon dioxide and the gases used in your refrigerator, indicate that sufficient pressure can cause them to liquefy at room temperature. However, certain gases, which became known as permanent gases, proved intractable to liquefaction through pressure alone. This was explained by the Dutch physicist Johannes van der Waals, who realized that intramolecular forces

could act to make liquefaction through pressure alone extremely difficult; the temperature would have to be lowered substantially in order
for pressure to produce liquefaction.

Reaching those lower temperatures required what became known as
the cascade approach to liquefaction, and it was to produce a breakthrough in the search for the lowest temperatures. The cascade approach involved liquefying one gas and using that gas to lower the
temperature of another gas, then pressurizing the second gas in order
to liquefy it. The first of the permanent gases to fall to this technique
was oxygen, and then nitrogen. Finally, the brilliant Scottish physicist
James Dewar successfully assaulted what he had called "Mount Hydrogen," achieving the liquefaction of that gas at about –420°F.[5]

Although it took Dewar more than a decade to accomplish his goal,
he was not to receive the accolades from the scientific community to
which he believed he was entitled. Unfortunately for Dewar, while he
was reaching the summit of Mount Hydrogen, Sir William Ramsay had
managed something even more amazing—he isolated helium, a gas that
had been presumed to exist only on the Sun. Imagine how Sir Edmund
Hillary would have felt if, after reaching the peak of Mount Everest,
he had seen another, higher mountaintop beckoning tantalizingly in the
distance, and then learned that someone else had already climbed it.

The man who was to beat Dewar in the race to liquefy helium, at a
bare 2 degrees above absolute zero, was the Dutch physicist Heike
Kamerlingh Onnes. Onnes and the British physicist Sir Ernest Rutherford, though working in different countries on different problems, simultaneously developed what is now known as "big science."
Generations of scientists had worked either as lone wolves or in small
groups, but Onnes and Rutherford created laboratories staffed by a
team of scientists and technicians. The race, as Damon Runyon would
put it, is not necessarily to the swift nor the battle to the strong, but
that's the way to bet.[6] Kamerlingh Onnes employed a variation of the
cascade technique, using oxygen to liquefy nitrogen, nitrogen to liquefy
hydrogen, and hydrogen to liquefy helium—the same approach that
Dewar was using, but with greater resources. Success was achieved in
June 1908.

This achievement was to have unexpected and exciting consequences. In cataloging the properties of liquid helium, Kamerlingh Onnes decided to measure its electrical resistance. Every other substance previously investigated had some electrical resistance, but, at a sufficiently low temperature, liquid helium had none; an electrical current induced in liquid helium will flow forever. This made it the first-known superconductor. (The search for high-temperature superconductors today is one of the major quests of contemporary physics; a material that is superconducting and easy to shape at room temperature would have substantial economic benefits.) Equally astounding was that liquid helium has no viscosity—the property of self-adhesion that makes honey and molasses so difficult to pour. Liquid helium placed in an open container appears to defy gravity; it rises of its own accord and overflows the container.[7]

One of the hallmarks of the progress of science and technology is how stuff that once was rare becomes widely available; what was the subject of an epic quest a century ago is now an item of basic commerce. The price of a liter of liquid helium is about the same as a latte and biscotti at your local Starbucks. Keeping it, however, is another matter—you'll probably have to store it in a special Dewar flask, and these go for thousands of dollars. It somehow seems fitting that both Dewar and Kamerlingh Onnes are remembered for their achievements in pursuing the ultimate in cold, even if they are remembered in different ways.

Bose-Einstein Condensates
and the Realm of the Ultimate Cold

I was born about a decade before room air conditioners came into widespread use. Summer nights in the suburbs of New York were liable to be hot and sticky, so we took advantage of a simple way of getting cool enough to sleep. You simply spread a thin film of water on your body just before going to sleep; the water molecules would evaporate and in so doing would remove heat from your body, cooling it off. The same phenomenon explains why a hot liquid left in an open cup cools; the molecules have an average heat but the hottest ones evaporate, leaving behind molecules of a lower average temperature.

An ingenious application of evaporative cooling to liquid helium enabled scientists of the mid–twentieth century to reach temperatures within one-thousandth of one degree of absolute zero. Scientists, however, wanted to go much farther than that—they were seeking a new state of matter, the existence of which had been predicted by Albert Einstein,[8] but which could only be found at temperatures within a hair's breadth of absolute zero.

The impetus for this arose in what, at the time, was an unlikely place: India. A century ago, India was off the beaten track as far as science was concerned. Nevertheless, shortly before World War I, an obscure Indian mathematician, Srinivasa Ramanujan, wrote a letter describing some interesting results he'd found to the Oxford mathematician G. H. Hardy. Hardy described that letter as the one truly romantic moment in his life. Hardy knew some of Ramanujan's results, suspected some of the others, and found some so surprising that, as he wrote, "They must be true, because if they were not true, no one would have the imagination to invent them."[9] Fittingly, the letter made Ramanujan an international mathematical star. A decade later, the Indian physicist Satyendra Bose wrote Albert Einstein a letter about the statistical mechanics of photons. Possibly remembering what had happened when Hardy received the letter from Ramanujan, Einstein read Bose's letter, and was so impressed that he translated Bose's results into German and submitted it on Bose's behalf to the prestigious *Zeitschrift für Physik*[10] ("Journal of Physics"). Einstein extended Bose's work to certain other particles, which resulted in the prediction of a state of matter that had not yet been shown to exist. This state of matter, called a Bose-Einstein condensate, could only occur at temperatures inconceivably close to absolute zero. A Bose-Einstein condensate consists of a collection of bosons (particles with integer spin, which can either be elementary, such as the particles that carry forces, or composite, such as the nucleus of a carbon-12 atom), all of which occupy the lowest possible quantum energy state. In such a state, these particles lose their individual identities; they are not just "all for one and one for all," but all are one and one is all.

Bose and Einstein had shown that it would require temperatures much colder than that achieved by evaporative cooling of liquid helium for a Bose-Einstein condensate to exist.

In order to achieve this, a new piece of technology was needed— the laser. As an atom cools down, its kinetic energy decreases. Since kinetic energy depends upon an atom's mass and velocity, this requires that its velocity has decreased as well. Absolute zero represents the temperature of an atom that does not move at all, which quantum mechanics has shown to be impossible. Nonetheless, the technique of laser cooling allows atoms to be slowed down to a speed that is nearly indistinguishable from zero. The idea is reasonably straightforward; if an atom is moving in one direction and collides with a photon moving in the other direction, the atom will absorb some of the energy from the photon. Just as hitting a tackler slows down a runner in football, hitting the photon will slow down the atom—as long as the photon has a frequency that "resonates," or is in synch with, the natural frequency of the photons that are characteristically emitted by the atom.

The race to produce a Bose-Einstein condensate paralleled the race to liquefy helium nearly a century earlier. The competition the second time was much more of a friendly one, with the competing teams meeting at conferences, exchanging notes, results, and ideas. One group was headed by Eric Cornell and Carl Wieman at the University of Colorado at Boulder, the other by Wolfgang Ketterle at MIT. Cornell and Wieman got there first, achieving a Bose-Einstein condensate in a cluster of approximately two thousand rubidium atoms cooled to less than a millionth of a degree above absolute zero. This was soon followed by a similar success by Ketterle with a much larger collection of atoms, and all three were awarded the Nobel Prize for Physics in 2001.[11]

I recall that, on my visit to the Hayden Planetarium, the docent mentioned that even though the temperature of the solar plasma was over a million degrees, you'd never feel it. I found this astounding—after all, I knew that if I were to spill some boiling water on myself, I would certainly feel it. However, the solar plasma is so incredibly thin that the heat content of the plasma is almost nonexistent. Similarly, you should absolutely never pick up a piece of dry ice (frozen carbon dioxide, about -110°F), but a simple calculation should convince you that you could pick up a Bose-Einstein condensate without any damage. The original Bose-Einstein condensate contained only 2,000 rubidium atoms. A grain of salt, on the other hand, contains approximately 10^{18} atoms, in a cube of 1 million atoms on a side. A cube of 2,000 atoms would have approximately 13 on a side. Admittedly, a rubidium atom is somewhat larger than a sodium or a chlorine atom (the atoms that make up salt), but it's probably safe to say that if the Bose-Einstein condensate of 2,000 rubidium atoms formed a cube, each side is less than $\frac{1}{10,000}$ of the length of one side of a grain of salt. Even at almost absolute zero, it seems pretty certain that you could pick it up safely— assuming you could even find it.

A Final Twist: Negative Temperature

The Fahrenheit and Celsius scales used for temperature measurement throughout the world offer negative temperatures because a tempera-ture of zero is merely a reference point. It's a cold day, but nothing ex-ceptional, if the temperature drops below zero degrees Celsius, and it's a very cold day if the temperature drops below zero degrees Fahrenheit, but anyone who lives in either the Midwest or Canada is used to this. However, absolute zero connotes an absence of movement, which is prohibited by quantum mechanics. If absolute zero corresponds to no movement, one would think that temperatures below absolute zero would correspond to quantities not of this universe. Maybe tachyons, hypothetical objects with imaginary mass that never travel slower than the speed of light, could have negative kelvin temperatures, but what could possibly be meant by less movement than no movement?

The problem here is that we are defining temperature in terms of motion, which is the approach of classical physics. Statistical mechan-ics offers a broader definition of temperature, however, which can make negative temperatures possible. A precise explanation requires an un-derstanding of both calculus and entropy, but it is possible to get an idea of how negative temperatures can occur without dealing too much with either, although we'll see a more detailed discussion of entropy when we discuss thermodynamics in Chapter 7.

Here's a familiar example that illustrates the idea of entropy. I recently attended a formal dinner party, in which there were two distinct periods: the predinner cocktails and the dinner itself. During the cocktail period, people were moving around freely, but once the dinner began, everyone went to their assigned seats. An informal definition of entropy is the number of different ways of arranging the individual components corresponding to a higher-level description of the system. There are only two higher-level states of the party: the cocktail period and the sit-down dinner. The individual components of the system are the guests. The cocktail period had higher entropy than the sit-down dinner, because there were many more arrangements of the individual parts (the guests) during the cocktail party than during the sit-down dinner. Statistical mechanics uses the term "macrostate" for the higher-level description of the system, and "microstate" for the lower-level one. A microstate corresponding to the cocktail period macrostate would read something like, "Fred was at the bar drinking a martini and talking to Anita, who had just ordered a Bloody Mary from the bartender."

Now consider a glass of water with an ice cube floating in it. Two adjacent ice molecules are constrained to be near each other and moving in roughly the same direction. When the ice cube melts, however, those two molecules are free to drift anywhere inside the glass and move unrelated to each other. The water and ice cube state has fewer arrangements of the individual molecules constituting it than does the glass of water because there is no "adjacent molecule" restriction on the water molecules as there is on the ice molecules.

It is possible for entropy to decrease, but in the ordinary world, we have to pump energy into the system to make that happen. Left to their own devices the guests at the party will mill around drinking cocktails; the hostess has to make an announcement that it's time to sit down for dinner—or lead by example by taking her seat. We can get a glass of water to morph into a glass of water with an ice cube floating in it, but we have to supply the energy to enable refrigeration. The fact that we can do this enables a more mathematically precise definition of temperature to be given—rather than it being a function of the kinetic energy of the particles, it is the rate at which entropy changes as more energy is added to the system ($T = dS / dE$ for those who are familiar with calculus).

Of course, for most systems with which we are familiar, adding energy will increase entropy. A system near absolute zero has all its molecules moving extremely slowly, which implies few microstates and therefore low entropy. Add heat to the system and the molecules move more quickly and are not constrained to be so close to one another, so the entropy increases. This positive change in the amount of entropy as energy is added to the system gives a positive value to the quantity dS / dE, and hence we always observe temperature as positive. That's because there are theoretically limitless microstates available to the system; heat it enough and we can imagine the molecules flying all over the place at incredible speeds. However, there are systems that have only a limited range of available microstates—and for those systems something happens that has no parallel in our everyday world. There comes a point when adding extra energy results in a decrease in entropy; that corresponds to a negative temperature because a decrease in entropy as energy is added to the system gives a negative value to the quantity dS / dE.

Quantum theory made it possible to envision such systems, and it's not hard to see how negative temperature (a decrease in entropy as energy is increased) can occur. Suppose we have four atoms confined in a very thin wire, perhaps confined by magnetic trapping, so that when energy is added, all the energy goes to changing the spin of the atom rather than its position or velocity. We'll assume this is not a Bose-Einstein condensate, so atoms 1, 2, 3, and 4 have separate identities. Each atom only has two states: the spin down (low energy) state, or the spin up (high energy) state. Imagine the system is initially in its lowest energy configuration with all four atoms having spin down; this is the only microstate associated with this energy configuration. If one quantum of energy is added, there are four possible microstates associated with this; any one of the four atoms could be spin up with all the others spin down. The number of microstates has increased, so the addition of energy has resulted in an increase in entropy; this corresponds to positive temperature. Add another quantum of energy and there are six possible corresponding microstates, depending on which two atoms have spin up (1 and 2, 1 and 3, 1 and 4, 2 and 3, 2 and 4, or 3 and 4). Again an increase in energy has resulted in an increase in entropy. However, add one more quantum of energy, and there are only four possible microstates, corresponding to which one of the four atoms has spin down while all the others have spin up. Here an increase in energy has resulted in a decrease in entropy, and so the temperature is negative—below "absolute zero."

Stranger Than We Can Imagine?

That's not the only bizarre situation that comes about with negative temperature. Bizarre though it may seem, a system at negative temperature is hotter than the same system at positive temperature; heat will flow from the system at negative temperature to the system at positive temperature! Heat always flows from hotter to colder systems—but this is not measured by temperature. In fact, the temperature scale from numbingly (and beyond) cold to blisteringly (and beyond) hot increases from just above 0°K to positive K to positive infinity K (although of course it can't get there), and then jumps to negative infinity K (likewise), negative K, to just below 0°K. Slightly below absolute zero is considerably hotter than hell.

The brilliant astrophysicist Sir Arthur Eddington once said that the universe is not only stranger than we imagine, it is stranger than we can imagine.[12] Maybe not—if we can come up with the mathematics to describe it. An argument might be made that negative temperatures are an artifact of the mathematical expression we use to define temperature, but that definition was motivated by defining temperature as a quantity that had the same value for any two systems in thermal equilibrium.

Negative temperatures don't seem to occur naturally anywhere in the universe, but we can look at the thermodynamical equations describing temperature, couple them with our knowledge of quantum mechanics, and predict phenomena that may be beyond even the universe's

power to imagine. Negative temperature systems have been produced
in the laboratory and have been studied for almost half a century—but
as yet nothing has been produced that has shown up in your local stores.
Just wait—a hundred years ago a few drops of liquid helium was the
subject of nearly two decades of intense effort; now you can find it on
Google Shopping for $5.00 per liter.[13]

CHAPTER 5

AVOGADRO'S NUMBER

Physics may lay claim to being the most fundamental of the sciences, but chemistry is the science that has the most influence on our lives—and more importantly, on the quality of our life. This is a book about basic science, but I'd like to pause for a moment to consider the economy. Given how much time we spend participating in the economy—in the production, distribution, and consumption of "goods and services," if you will—you might think that economics is the most important science around. It describes our jobs, after all, and taxes, and most everything that takes up way too much of our time. But you can't have economics without goods and services, and where do these goods and services come from? Moreover, services are clearly a second fiddle in that duo, given that, without goods, there wouldn't be a whole lot in the way of services. After all, you never hear anyone say, "He really delivers the services." And when it comes to delivering the goods, you simply can't beat chemistry. I've seen estimates that 10 billion different items are available for consumption in London alone,[1] and a substantial fraction of recently developed goods rely heavily on our knowledge of chemistry for their existence. Fundamental to that knowledge is our target in this chapter: Avogadro's number, which tells us the number of particles that exist in a certain amount of stuff. Without it, chemistry would probably still be the hit-or-miss operation it was when alchemists mixed potions in the hope that something interesting would occur, and the multitude of goods that make modern life so enjoyable would never have appeared.

Better Things for Better Living

I was born shortly after the 1939 World's Fair, which took place during the spring and summer just before the invasion of Poland that marked the beginning of World War II. In retrospect, we can look with some wistfulness at the fair and its promise: that we were seeing the "World of Tomorrow," the way the world could be. This is not to say that everything at the fair turned out to be nothing but daydreams, however. One section of the fair was the Production and Distribution Zone, and one of the chief exhibitors there was the Du Pont Company, one of the world's foremost chemical companies. Featured were two of Du Pont's synthetic creations, neoprene and nylon, the two spearheads in what would eventually become the plastic takeover of the world. These had been synthesized by a team headed by Wallace Carothers,[2] a chemist who had been enticed to move from Harvard to Du Pont because of the extensive research facilities available in industry.

Carothers's—and Du Pont's—impact has been tremendous. I work sitting at my desk, constructed of some sort of plastic, typing on a keyboard that is primarily plastic, and visualizing the words I type on a monitor that is also mostly plastic. I took an inventory of the items on my desktop: there are at least twenty separate items whose components are largely plastic. My shoes are mostly plastic, as are the rims of my eyeglasses. Suffice to say that the whole room is practically a shrine to the benefits of synthetic organic chemistry, which makes it an ironic tragedy that Wallace Carothers was born too early to benefit from a later triumph of chemical synthesis. Carothers suffered from severe mood disorders, and committed suicide two decades before the widespread introduction of the antidepressants that have helped so many afflicted with this condition.

The first great step on the road to creating the World of Tomorrow—and yesterday's world of tomorrow is now the world of today—was taken by John Dalton, whom we have encountered before. In the chapter on ideal gases, we looked at one of Dalton's failures—his belief that gases expanded in proportion to their previous volume. Many great scientists have a notable failure on their record—Linus Pauling proposed a triple-helix structure for DNA, missing it by *that much*—but it is their successes that characterize their careers, and Dalton's exposition of the atomic theory is one of the great landmarks of science. Without the atomic theory, chemistry would be basically a hit-or-miss collection of cookbook recipes. Without the atomic theory, there is simply no way to deliver the goods in the abundance that we now enjoy.

The basic principle of Dalton's atomic theory was, first and foremost, that all matter is composed of extremely small particles, called atoms. This makes chemistry the study of how the atoms of substances combine and separate to create different substances. According to Dalton's theory, each element has its own characteristic atom and each compound its own characteristic molecule (Dalton called them "ultimate particles"), and all examples of any particular particle are identical. Finally, when elements combine to form compounds, the molecules of the compounds consist of small whole numbers of atoms of the constituent elements. Although modern science has unearthed and created exceptions to these basic premises, they are honored much more in the observance than in the breach. Indeed, as Richard Feynman put it in an introductory lecture at Caltech in 1961, "If, in some cataclysm, all of scientific knowledge were to be destroyed, and only one sentence passed on to the next generation of creatures, what statement would contain the most information in the fewest words? I believe it is the atomic hypothesis . . . that all things are made of atoms—little particles that move around in perpetual motion. . . . "[3]

But it was a long and rocky road from Dalton and the atomic theory to nylon, Prozac, and the transformation of the 1939 World Fair's World of Tomorrow into the world of today. The atomic theory was the first critical step, but simply knowing the nature of the bricks doesn't guarantee you can build a building. A great deal needed to be accomplished before the Du Pont Company could embody its slogan "Better Things for Better Living—Through Chemistry."

The Structure of Chemical Compounds

Bricks, when put together, make buildings—and atoms, when put together, make chemical compounds. Just as the knowledge of how many bricks are needed and where they should be placed is necessary to construct a building, it is necessary to know what atoms are needed and where they should be placed to create chemical compounds. Admittedly, in the early portion of the nineteenth century, scientists were not aware of the role that the arrangement of the atoms within a molecule played; simply determining the number of atoms in a compound was a difficult task.

Dalton's original theory allowed for the construction of different compounds using the same elements. For instance, Dalton knew of two distinct oxides of carbon, what we nowadays refer to as carbon monoxide and carbon dioxide. He was able to ascertain that a carbon monoxide molecule consisted of one atom of carbon and one of oxygen, whereas a carbon dioxide molecule required two atoms of oxygen to accompany the one atom of carbon. This enabled him to deduce the relative weights of the carbon and oxygen atoms with some accuracy. This accuracy, however, was somewhat impaired by Dalton's incomplete knowledge of the chemical composition of that most vital and ubiquitous substance—water.

Dalton had developed a systematic approach to the description of chemical compounds. In Dalton's scheme, a binary compound was one in which a single atom of one element was paired with a single atom of another element; a ternary compound was one in which a single atom of one element was combined with two atoms of another element; a quaternary compound, a single atom of one element combined with three atoms of another element. This scheme worked well to describe the binary compound carbon monoxide and the ternary compound carbon dioxide. However, in Dalton's era only one compound of hydrogen and oxygen was known—water—and so Dalton assumed it consisted of a single atom of oxygen and a single atom of hydrogen. Dalton knew that the weight of the water formed from equal volumes of oxygen and hydrogen was nine times the weight of the hydrogen component, and because he assumed that the same number of hydrogen and oxygen atoms were present in water, he deduced using simple algebra that the weight of an oxygen atom was approximately eight times the weight of a hydrogen atom.

Amedeo Avogadro and Stanislao Cannizzaro

The resolution of the problem of relative atomic weights was to come from the efforts of two Italian scientists. Ironically, Avogadro is not universally regarded as a chemist, given the pivotal role that Avogadro played in the development of modern chemistry. True, Avogadro was a professor of mathematics and physics throughout most of his career, but his fame rests on a single paper he published in 1811, with the exhausting title "On the Determination of Proportion in Which Bodies Combine According to the Number and the Respective Disposition of the Molecules by Which Their Integral Particles Are Made." Publicity was harder to get in the early nineteenth century than it is today—no Internet on which to blog, no search engine to game—and so this paper with its leaden title sank from sight. Too bad, because Avogadro had developed a rationale that could have hastened the development of chemistry by allowing chemists to correctly compute the relative atomic weights of the elements.

That rationale is contained in Avogadro's hypothesis, which is very simple to state: at the same temperature and pressure, equal volumes of gas contain the same number of molecules—*not* the same number of atoms. The distinction between atoms and molecules was not completely clear to Dalton and the other scientists of that era. Dalton recognized that the molecule was the basic indivisible unit for chemical compounds, but he did not realize that for some elements, such as hydrogen, nitrogen, and oxygen, the atoms of the same element would form bonds with each other, giving us diatomic molecules of hydrogen, nitrogen, and oxygen, for example. (Indeed, more generally, he thought of atoms as indivisible particles, something not definitively disproven until Otto Hahn and Fritz Strassmann initiated the fission of uranium into barium in 1938.) We have seen how his erroneous structure for the water molecule resulted in an inaccurate determination for the atomic weight of oxygen. This problem was corrected by Avogadro's hypothesis and the methodology it suggested.

The critical experiment involves combining two liters (or any fixed volume) of hydrogen with one liter of oxygen; so long as both those volumes are at equal temperatures and pressures, what results is two liters of water vapor at the same temperature and pressure. Avogadro's hypothesis is that equal volumes of gas contain the same number of molecules; consequently, there are the same number of water molecules at the end as there were hydrogen molecules at the beginning, and there are twice as many water molecules as there were oxygen molecules. Therefore, a single molecule of water contains twice as many hydrogen atoms as it does oxygen atoms, which means that the simplest form for a water molecule is the well-known H_2O. This enables one to compute the correct atomic weight for oxygen, which is not eight times that of hydrogen, as Dalton surmised, but nearly sixteen times. Indeed, Avogadro actually made this computation, getting a weight of fifteen.

There is another, subtler deduction that can be made from Avogadro's law. If the volumes of hydrogen and oxygen were collections of single atoms rather than diatomic molecules, the chemical reaction described above would produce the same number of water molecules as there were oxygen atoms—that is, there would be one liter of water vapor, not two. The way out of this dilemma was to assume that each molecule of hydrogen consisted of two atoms of hydrogen and each molecule of oxygen consisted of two atoms of oxygen. Then two molecules of hydrogen would consist of four atoms of hydrogen, one molecule of oxygen would consist of two atoms of oxygen, and the resulting reaction would produce two molecules of water. Each water molecule would consist of two hydrogen and one oxygen atoms, and the two water molecules would consist of four hydrogen and two oxygen atoms, a correct job of atomic bookkeeping.

Avogadro was fully aware of the consequences of his hypothesis. As he states in his seminal article, "Setting out from this hypothesis, it is apparent that we have the means of determining very easily the relative masses of the molecules of substances obtainable in the gaseous state, and the relative number of these molecules in compounds; for the ratios of the masses of the molecules are then the same as those of the densities of the different gases at equal temperature and pressure, and the relative number of molecules in a compound is given at once by the ratio of the volumes of the gases that form it.[4]

The history of science is replete with advances that went unrecognized when they were published, and Avogadro's contribution remained in the background for almost half a century. The classic four-volume recapitulation *History of Chemistry*, published between 1843 and 1847 by the German chemist and historian Hermann Kopp, contains no reference to it. Indeed, when Avogadro died in 1856, his obituary notice in the journal *Nuovo Cimento* did not mention it! It took the efforts of a fellow Italian, Stanislao Cannizzaro, to bring attention to the numerous problems that Avogadro's hypothesis resolved.

Perhaps Avogadro's hypothesis was ahead of its time, and chemists of the era were not yet sufficiently comfortable analyzing chemistry in terms of atoms and molecules to recognize its power. Perhaps it was just that he was not considered a chemist by his peers. Cannizzaro, on the other hand, was a chemist of some stature. He had discovered an unusual type of reaction, now known as a Cannizarro reaction,[5] so perhaps his colleagues were just more prepared to listen to him. (The reaction is beyond the scope of this book, but it involves—for those readers familiar with chemistry—both reducing and oxidizing one chemical species to get two different chemical species.) Regardless, I do wonder how chemistry could have advanced to a point where an analysis of such a complicated reaction was possible without taking advantage of the clarity and power of Avogadro's hypothesis.

Two years after Avogadro's obituary in *Nuovo Cimento* omitted mention of the hypothesis, Cannizzaro published an article in that journal in which he restated the hypothesis and discussed how it resolved many of the perplexing problems that chemists were encountering. This time, the chemical world was ready to appreciate it. As the great German chemist Julius Lothar Meyer later wrote, "Avogadro's hypothesis had a great influence particularly upon the development of chemical theories. . . . From Avogadro's laws dates the beginning of a general theory of chemistry, a theory that explains the atomic constitution and the major part of the properties of compound bodies."[6]

Dmitri Mendeleyev and the Periodic Table of the Elements

By the time Cannizzaro made Avogadro famous, the world's chemists had discovered sixty-three elements. Even though they now had a tool for determining the atomic constituents of compounds, they still lacked general rules for describing what various atoms, and the compounds they could form, were like. For example, when sodium, a lightweight, explosive metal, chemically combined with chlorine, a poisonous yellow-green gas, the result was common table salt, sodium chloride, a compound that is neither metallic, gassy, poisonous, or explosive. Until the rules could be discovered, the potential of chemistry would be limited.

Into this state of disarray came Dmitri Mendeleyev, a Russian chemist, who decided to try to organize the known elements into a pattern. To do so, he first arranged these elements in increasing order of atomic weight, the same physical property that had attracted the attention of John Dalton when he devised the atomic theory. He then imposed another level of order by grouping the elements according to secondary properties such as metallicity and chemical reactivity—the ease with which elements combined with other elements.

The result of Mendeleyev's deliberations was the first periodic table of the elements, a tabular arrangement of the elements in both rows and columns. The atomic weights increased from left to right in each row, and from top to bottom in each column, and, in essence, each column was characterized by a specific chemical property—alkaline metals in one, chemically nonreactive gases in another.

When Mendeleyev began his work, not all the elements were known. As a result, there were occasional gaps in the periodic table: places where Mendeleyev would have expected an element with a particular atomic weight and chemical properties to be, but no such element was known to exist. With supreme confidence, Mendeleyev predicted the future discovery of three such elements, giving their approximate atomic weights and chemical properties even before their existence could be substantiated. His most famous prediction involved an element that Mendeleyev called eka-silicon. Located between silicon and tin in one of his columns, Mendeleyev predicted that it would be a metal with properties resembling those of silicon and tin. Further, he made several quantifiable predictions: its density would be 5.5 times greater than that of water, its oxide would be 4.7 times denser than water, it would be gray, and more. When eka-silicon (later called germanium) was discovered some twenty years later, Mendeleyev's predictions were right on the money.

In addition to being one of the great organizing principles of science, the periodic table has tremendous practical importance. If a compound is useful but has some undesirable properties due to one of its constituent atoms, it may be possible to find a better compound for the same purpose by substituting a chemically similar atom in the place of the problematic one. For those who must regulate their sodium intake, an acceptable alternative is so-called light table salt, in which potassium chloride replaces sodium chloride. The taste is similar, but its effect on our blood pressure is not.

Scientists often develop their theories in surprising fashion. It was necessary for Mendeleyev to engage in countless restructurings of his periodic table, as he had no idea at the start how many rows and columns would be required. To write down the results of each trial would tax anyone's patience. So Mendeleyev constructed a deck of cards in which each card contained the name and properties of a specific element. Playing solitaire with this deck of cards made it easier and more entertaining to try the different possibilities for the periodic table. (Appropriately enough, the nineteenth-century name for a version of solitaire was Patience.)

Avogadro's Number

Taken together, Avogadro's hypothesis, Dalton's atomic theory, and Mendeleyev's periodic table form much of the bedrock of chemical science, but not all of it. Avogadro provided one other important piece of that foundation when he deduced Avogadro's number—these days, perhaps better known as the Avogadro constant, which describes the number of particles in one mole of a substance. A mole is defined as the number of particles in 12 grams of carbon.

Avogadro's initial formulation predicted that two equal volumes of two different gases would contain equal numbers of particles. The masses of those gases, however, are not the same, and Avogadro's number enables us to numerically connect those two measurements, mass and volume. The mass of a liter of carbon gas would be twelve times that of a liter of (monatomic) hydrogen gas, and 25 percent less than that of (monatomic) oxygen. The most recent determination of the number of particles in Avogadro's number is $6.02214179 \times 10^{23}$.[7]

The obvious question is, how do we know this? I suppose the most direct approach would be to take one mole of a gas and simply count the number of molecules. Of course, this is impossible—at least for the present—so another method has to be found. Amazingly enough, it turns out that there are a lot of ways to do this, but modern technology has made the problem relatively straightforward. One takes a material whose crystalline structure is cubical (silicon is the current choice), so that the atoms in the crystal are the same distance apart whether one goes from an atom to the next atom by going "north," "east," or "up" (or equivalently "south," "west," or "down"). Modern lasers can measure the distance between atoms with considerable accuracy, and the number of atoms in a mole can then be calculated in a fairly straightforward fashion. It's basically the same way one would calculate the number of trees in an orchard, provided that they were planted in rows and columns and one knew the distance between adjoining trees in the same row or the same column.

However, lasers have only been around for fifty years or so. The first time I saw one was during the classic bit of repartee in the movie *Goldfinger*, when Auric Goldfinger trains a laser on a metal block to which James Bond is chained. The laser burns ever closer to Bond, who asks Goldfinger, "Do you expect me to talk?"

"No, Mr. Bond," Goldfinger replies, "I expect you to die." Not surprisingly, Bond didn't die, and perhaps more surprisingly, scientists were able to do a pretty reasonable job of calculating Avogadro's number long before the advent of lasers.

John Strutt (Lord Rayleigh) performed a simple but ingenious experiment which gave an idea of the magnitude of Avogadro's number. He put one milligram of oil on a water surface and let it spread out. When it did, it covered a surface whose area was measured to be 0.9 square meters, or 9,000 square centimeters. The density of the oil was 0.9 grams per cubic centimeter, and the atomic weight of the oil was 282.5. Since the oil spreads out until it can no longer do so, the resulting oil slick is one molecule thick, and the volume of the oil slick is the height h of a single molecule multiplied by the area of the oil slick. The volume of the oil slick is the same as the volume of the original milligram of oil. Density is mass divided by volume, so $0.9 = 0.001$ grams / volume, and we see that the volume of the oil slick is $0.001 / 0.9 \approx .00111$ cubic centimeters.

Since the volume of the oil slick is 1.111×10^{-3} cubic centimeters, and we can think of it as a cylinder whose base is 9,000 square centimeters and whose height is h, we see that $9,000 \times h = 1.111 \times 10^{-3}$. So $h = 1.111 \times 10^{-3} / 9,000 = 1.234 \times 10^{-7}$ centimeters. Assuming that the space a single molecule occupies is a cube of side h (even though the molecule may not be shaped like a cube, that's the space it occupies, much like a cubical box is required to pack a basketball), we can estimate the number of molecules in the milligram of oil. The volume of that milligram was 1.111×10^{-3} cubic centimeters, and so the number of molecules in a milligram is approximately $1.111 \times 10^{-3} / h^3$. The number of grams in a mole of oil is the same as its atomic weight, 282.5, so the number of milligrams in a mole of oil is $282.5 \times 1,000 = 282,500$. Consequently, the number of molecules in a mole of oil—which would be Avogadro's number—is $282,500 \times 1.111 \times 10^{-3} / (1.234 \times 10^{-7})^3 = 1.67 \times 10^{23}$. While this is off by a factor of four from the current best estimate, it's still in the ballpark, which is what we would hope for from such a rough calculation with many estimates.

There is an underlying aesthetic in certain formulas, a classic example being Euler's Formula: $e^{i\pi} + 1 = 0$. You'd have to have a complete absence of aesthetic sense not to appreciate this formula—and of course I'm not worried about your lack of an aesthetic sense, as you are reading this book. This exquisitely beautiful formula unites the base of the natural logarithms, the ratio of the circumference of a circle to its diameter, the fundamental imaginary number, and the additive and multiplicative identities all in one glorious expression. It's like going to a rock concert and finding that the show consists of Elvis, Bruce Springsteen, the Rolling Stones, the Beach Boys, and Abba—maybe not *your* five greatest rock acts, but mine, and you can alter this list to suit your tastes. Avogadro's number has a similar power to Euler's famous formula, and we will see it again throughout this book.

How Large Is 6×10^{23}?

We don't encounter numbers this large in everyday life. The current national debt is almost \$14 trillion,[8] which is 1.4×10^{16} dollars. If there were approximately 43 million countries, each with that amount of national debt, it would total about 6×10^{23} trillion dollars, but it's impossible to imagine that many countries with the productivity of the United States without getting into the realm of interstellar civilizations. It's also impossible to imagine the process by which such a far-flung interstellar civilization could have evolved while simultaneously running up such a massive amount of debt.

One of the analogies I read when I was young that was used to illustrate the size of Avogadro's number was to take an ordinary cup of coffee and throw it into the ocean. Mix the oceans of the world thoroughly, and then fill the original coffee cup with ocean water. The cupful of ocean water will contain a few molecules of the original coffee, because the ratio of the volume of the cup to the volume of the oceans is on the order of Avogadro's number.

Here's a more modern way to look at it. I have a fairly new, but fairly cheap, computer—I recently ran a timing cycle on it and it can go through a loop a million times in roughly two seconds (I just timed it). I first started programming computers in the late 1950s—when computers were bulky, slow, and expensive. Only companies could afford to purchase them, so the fact that all this computer power sits on my desk, and cost well under $1,000 (including peripherals) is simply astonishing. A slightly faster machine—or my machine with programming dedicated to computation—would run at about a million loops per second, so let's assume that we have a machine that counts molecules at the rate of a million a second, and we give it the task of counting the number of molecules in a mole of an ideal gas. The universe is approximately 14 billion years old, and 14 billion years is $1.4 \times 10^{10} \times 365 \times 24 \times 60 \times 60 = 4.4 \times 10^{17}$ seconds, so assuming we started counting at the time of the big bang, it has counted approximately 4.4×10^{23} molecules. So it still has about another five billion years to run before it finishes.

CHAPTER 6

ELECTRICITY AND THE PROPORTIONALITY CONSTANT

M y first close encounter with electricity could easily have been my last.

At any rate, that's the way my mother told it, and since she's no longer here to confirm or deny, I'm willing to accept it as recounted to me. When I was about three years old or so, my mother noticed that I was diligently trying to insert a hairpin into an electric light socket. At any rate, she sprang into action—action consisting of telling me at maximum volume to stop doing that *and* frightening me sufficiently with regard to the potential dangers from electricity that to this day I never undertake any sort of complicated rearrangement of electrical connections without first donning rubber sneakers and work gloves. Everyone I know thinks this is utterly ridiculous, but if there is a fifth force in addition to gravitation, electromagnetism, and the strong and weak nuclear forces, it has to be maternal imprinting. I'm not sure how it compares with the strong nuclear force, which is pound-for-pound the strongest of the four forces, but in my case it's certainly stronger than electromagnetism—and in this chapter we'll see just how strong electromagnetism is.

Electricity and Magnetism—
The Early Years

Well, the early years as far as mankind's involvement with these forces is concerned—the early years of E & M date back to the big bang, or shortly thereafter. Both electricity and magnetism were known long before the birth of Christ. The Greeks knew that amber, which is fossilized pine resin, had the unusual property that, when rubbed, it could attract small pieces of wool and lint—and then suddenly expel them. Indeed, the very word *electricity* comes from *elektron*, the Greek word for amber. Lodestones—naturally magnetized pieces of magnetite, an iron ore—have been known even longer. When a lodestone is used to stroke a small piece of iron, the iron acquires the magnetic properties of the lodestone. In particular, iron needles that have been so magnetized will naturally align themselves in a north-south direction if allowed to do so. It was this property of magnetism that allowed the construction of an extremely significant invention—the magnetic compass.

The first magnetic compasses were probably invented by the Chinese more than two thousand years ago, and they were certainly being used in Europe for navigation by the twelfth century. For centuries, magnetic compasses were relatively simple affairs, consisting of a magnetized needle mounted on a sharp pin, the base of which was stuck through a card on which were marked the major compass points (N, E, S, and W) and some subdivisions (NW, SSW, etc.). The needle could swing freely, but since it always aligned itself in a north-south direction, the compass point N on the card could be placed under the needle, and directions determined with a fair degree of accuracy.

It would be a long time before any widespread use of electricity would be made. The key difference was that electric phenomena were transient, whereas magnetic phenomena were permanent, or nearly so. Not only did the longer duration time of magnetic phenomena enable them to find uses, it also enabled them to be studied more easily. The first person to investigate them seriously was William Gilbert.

William Gilbert

The term *Renaissance man* is used nowadays quite cavalierly to convey abilities in two relatively disparate areas. Many of the great thinkers of the fourteenth through seventeenth centuries were indeed Renaissance men; the archetype is probably Leonardo da Vinci. William Gilbert was another. He may not be nearly as well known as da Vinci, but his contributions to the advancement of science were even more significant.

Gilbert was born in the sixteenth century, an epic period in science. Vesalius had just published a groundbreaking book on anatomy, and the Copernican view of the universe was creating an intellectual firestorm. Unfortunately for some of its adherents, such as Giordano Bruno, a physical conflagration occurred as well—Bruno was burned at the stake for his heretical views. Ideas were more freely explored in England, where Gilbert had become the Royal Physician and was an early exponent of the Copernican theory.

In addition to practicing medicine, Gilbert was extremely interested in magnetic and electrical phenomena, and published the first great work on this subject, *On the Magnet and Magnetic Bodies, and on the Great Magnet the Earth*. In it, Gilbert correctly argued that the Earth was a giant magnet, which explained the fact that magnetized needles always aligned themselves in a north-south direction. Gilbert also described the electric force, which he termed *vis electrica*, and actually devised the first electric instrument, which used a pivoting needle—borrowed from the compass—to measure the relative abilities of various substances to be attracted by the electric force. As Gilbert wrote, "The electric effluvia differ much from air, and as air is the earth's effluvium, so electric bodies have their own distinctive effluvia; and each peculiar effluvium has its own individual power of leading to union, its own movement to its origin, to its fount, and to the body emitting the effluvium."[1]

Gilbert also noticed some important differences between the behavior of magnetism and the behavior of electricity. He noted that electrical attraction disappeared with heat, whereas magnetic attraction did not, and so concluded that the two forces were different. This is not actually true—an object's magnetic properties can be destroyed by extremely high heat, for one thing, and more importantly, as was to be discovered, the two phenomena are actually manifestations of the same force—but Gilbert's study was an important step toward creating a civilization that could harness electrical power. Unfortunately, it would be more than two centuries after Gilbert's death before the next step was taken.

Charge Accounts

To study electricity required that scientists be able to produce it more reliably and in greater quantities than Gilbert could—and to be able to store it. The latter difficulty had already been surmounted—although not by humans—as certain types of eels and fish are capable of delivering a nasty electrical shock. The first reliable and human-made source of a significant amount of electricity was invented in the seventeenth century by Otto von Guericke. Von Guericke also deserves to be called a Renaissance man; after studying mathematics, law, and engineering, he was forced to flee his home town of Magdeburg during the Thirty Years' War. On his return, he helped to rebuild the city, and was later elected its mayor—and then performed two notable scientific experiments. He invented a device, the Magdeburg hemispheres, to demonstrate the existence and power of the vacuum, and also devised the first large-scale electrostatic generator. This consisted of a large ball of sulfur with a rod as a central axis, which was surrounded by a belt. The sulfur ball could be turned with a crank and rubbed with dry hands, generating an electric charge on the sulfur ball, which could then be taken elsewhere and studied.

Despite von Guericke's work, it was only during the eighteenth century that the pace of electrical experimentation and development began to quicken. The first of the significant inventions was the Leyden jar, a device for storing electricity. Leyden jars could store significant amounts of electricity, enough to kill small animals. Leyden jars could be connected together—nowadays, we would say "in series"—in order to increase the amount of charge that could be conveyed when the electricity was released. The Leyden jar could also be used, as Franklin so famously demonstrated, to show that lightning itself was a form of electricity. Obviously, the power of lightning was well known, and once it was known conclusively that it was a form of electricity, a way was devised to tap lightning as a source of electricity to charge as many as fifty Leyden jars simultaneously.

The other great development was a different way to obtain electricity—easier than rubbing something with something else, and far less dangerous than trying to catch lightning in a bottle (a metaphor for us, an actual endeavor for Benjamin Franklin and others). Scientific advances are sometimes the result of serendipities, and one such occurrence took place when a scalpel that was in contact with a static electricity generator in the laboratory of Luigi Galvani accidentally touched the leg muscles of a frog that lay nearby. The muscles twitched, and Galvani began an extensive investigation of what he called animal electricity, which he believed was a candidate for the elusive "life force" that both philosophers and scientists had been hunting. Serendipity struck again when he bound the frogs' legs with copper wire and attached them to an iron balustrade around his terrace. The frogs' legs twitched—as they had before when touching the electrified scalpel—but this time there was no obvious source of electricity. Galvani then found that the twitching did not take place if the wires were made of the same metal as the balustrade, and thus discovered that electricity could be produced by bringing different materials together to generate it.

This development was brought to fruition by one of the great names in the history of electricity: Alessandro Volta. Volta began to systematically explore the idea that electricity could be produced through the physical contact of different metals. A careful experimenter, Volta tested many metals, producing an ordered list such that each metal would generate a positive charge when paired with a metal above it on the list. He had also noticed that, when he put two different metals in his mouth, his tongue tingled from the passage of the electric current, and so it occurred to him that moistening materials with brine might encourage the flow of electricity from metal to metal. He then stacked an array of differing metals separated by moistened cardboard to amplify the small electrical current produced by each metal pair. The result was the Voltaic pile, and it was not so different from the batteries you can buy today in your neighborhood store.

Now, finally, science had a source of electricity that was both reliable and continuous. The problem with Leyden jars was that, like a basketball team that loses in the first round of the NCAA tournament, they were "one and done"; they discharged their complete stock of electrical charge all at once. While this could be used for various parlor tricks and also to ignite explosives (a function that electricity is still fulfilling), it greatly limited the potential applications for electrical power. So, even while the charge available from the initial Voltaic pile was minute in comparison with the amount of electricity that could be stored in a Leyden jar, it was the opening salvo in the arms race to produce and use electricity.

Volta wrote a letter to Sir Joseph Banks of the Royal Society, in which he outlined the development of the Voltaic pile, and his conclusions with regard to it. This letter was read to the Royal Society on June 26, 1800—a date that serves to mark the formal beginning of the development of usable electric power. However, the source of the electric power from the Voltaic pile was not understood by Volta, and would not be fully appreciated for decades. Batteries work as the result of the release of electrons through chemical action, and are an example of both the law of conservation of energy and the laws of thermodynamics. The law of conservation of energy—which would later be amended to incorporate the equivalence between mass and energy in the iconic formula $E = mc^2$ discovered by Einstein—states that energy can neither be created nor destroyed. Consequently, the energy that emerged from the Voltaic pile in the form of electricity had to come from *somewhere*. The laws of thermodynamics explained that energy could be transformed from one form to another, but in so doing a certain amount of energy was inevitably lost. Like a currency broker that will exchange your dollars for euros but charge you a fee for doing so, the universe—

as described by the second law of thermodynamics—charges a fee (measured in lost energy) for transforming chemical energy into electrical energy.

Charles-Augustin de Coulomb

At the same time as the experimenters were looking for new ways to store and generate electricity, others were trying to explain and understand it. Fortunately, there was already a gold standard in existence for physical theories—Newton's universal law of gravitation. Newton had declared that the gravitational force between two bodies was directly proportional to their masses and inversely proportional to the square of the distance between them. This formula could have been the result of two basic hypotheses. The first would have been that if one of the masses doubles, the gravitational force between the two masses doubles as well. The second would have been that whatever gravity is, since it emanates from a point, it does so by spreading out over the surface of a sphere. The surface area of a sphere is proportional to the square of the radius, so if the distance between two masses doubles, the surface area, which surrounds the stuff that causes gravity, increases by a factor of four, diluting the stuff by that factor of four.

I'm displaying a good deal of hubris here in postulating how Newton might have thought, but it's a reasonably sensible approach, and if it occurred to me it could certainly have occurred to others. At any rate, electricity and magnetism could be subject to the same two hypotheses: double the stuff (electric or magnetic charge) of one of the objects and the force doubles, double the distance between them and the force decreases by a factor of four.

The man who successfully carried out the necessary experiments to demonstrate these laws for both electricity and magnetism was Charles-Augustin de Coulomb, an eighteenth-century French physicist. Coulomb's primary investigative tool was one we have seen before: the torsion balance devised by John Michell and used by Henry Cavendish to weigh the Earth. Coulomb's task was considerably easier than Cavendish's, however, because electromagnetic force is vastly stronger than the gravitational force. It's so much stronger, in fact, that the relevant deflections of the torsion balance from very small amounts of electric or magnetic charge can easily be measured in a laboratory, even a high-school one.

Coulomb wrote numerous memoirs on his investigations of the strength of the electric and magnetic forces. Here is an example from his *First Memoir*,[2] in which he introduced a like charge on two pith balls and measured their separation. I've greatly simplified the relevant data.

DETERMINATION OF THE FORCE LAW
FOR REPULSIVE (LIKE) CHARGES

INITIAL SEPARATION OF PITH BALLS	DEFLECTION
36.0	36.0
18.0	144.0
8.5	575.5

Notice that the initial separation halves (or nearly so) in each subsequent row, and the deflection quadruples, just as an inverse square law would predict. Writing nearly a century later, James Clerk Maxwell (who would become the ultimate arbiter of all things electromagnetic) said of Coulomb that "it is impossible to overestimate the delicacy and ingenuity of his apparatus, the accuracy of his observations, and the sound scientific method of his researches."[3]

In fact, there is a strong connection between the work of Michell and Cavendish, and the investigations of Coulomb. Coulomb's initial interest was in torsion, and in all probability the torsion balance that was built by Michell and used by Cavendish had been designed by Coulomb. Cavendish credits Coulomb in the *Philosophical Transactions* in 1798. "Many years ago, the Rev. John Michell . . . contrived a method of determining the density of the Earth . . . but, as he was engaged in other pursuits, he did not complete the apparatus till a short time before his death. . . . Mr. Coulomb has, in a variety of cases, used a contrivance of this kind for trying small attractions. . . . "[4]

Why Are Electricity and Gravity So Different?

The force laws for gravity and electricity look virtually the same when expressed mathematically. Newton's law of gravitation is $F = GmM / r^2$, where m and M are the two masses, r the distance between them, and G the gravitational constant. Coulomb's Law is $F = kqQ / r^2$, where q and Q are the two charges, r the distance between them, and k a constant of proportionality whose exact meaning and value we will see in a few pages. There are some important differences between the phenomena these two formulas describe. The masses m and M can have only positive values, and gravitational force is always attractive; every mass attracts every other mass. The charges q and Q, on the other hand, can take both positive and negative values, and the electric force is sometimes attractive (when q and Q have the opposite sign) and sometimes repulsive (when q and Q have the same sign).

An upshot of these two facts is that every one of us is a gravitational attractor, because we all have positive mass, but every one of us is electrically neutral, neither attracting nor repelling electrified objects (except on a cold day when we have walked across a rug and acquired a static electrical charge)—because the average electric charge of the particles in our body is zero. However, the theories of gravitation and electricity predict different behaviors for objects that have net gravitational charges (aka masses, and that's everything in the universe) and net electrical charges. Gravitational charges attract each other, and as Newton showed, any object has a center of gravity and the gravitational force emanating from that object can be regarded as emanating from its center of gravity. Additionally, masses attract each other, which explains why the molten iron core of the Earth is at the core of the Earth rather than making life impossible by covering the entire planet with a liquid iron ocean.

However, Coulomb showed that because like electrical charges repel each other, any net electrical charge that an object possesses will do its best to get as far away from the other electrical charges on the object as possible. Consequently, net electrical charges distribute themselves on the surface of an object. Although Coulomb had noticed this during his experiments, he proved it in a theorem in his *Fourth Memoir*.[5] For those who don't believe that mathematical theorems have practical consequences, here's some potentially lifesaving advice that is a consequence of Coulomb's theorem: if you happen to be caught on the road during a lightning storm, *stay in your car!* Even if your car is struck by lightning, thus acquiring a net electric charge, the charge will distribute itself on the outside of your car, and as long as you stay inside and don't touch anything electrically conductive that is connected to the outside of your car, you'll be fine. This was dramatically demonstrated by Nikola Tesla, who sat inside a "lightning cage" calmly reading while huge bolts of man-made electricity flashed all around him.

Hans Oersted, Michael Faraday, and Electromagnetic Induction

Any history of electricity, even so brief a one as contained in this chapter, would be incomplete without mentioning the two key experiments that probably did more to change the standard of living of the average human being than any other event in human history. The first of these two experiments was performed by the Danish physicist Hans Oersted, who in 1820 had shown that turning an electric current on and off near a compass would cause the compass needle to move. The compass needle normally only moved in response to the presence of a magnet; consequently, turning the electric current on and off generated a magnetic field. Of course, the word *field* had not yet been devised to describe how magnetism or electricity worked; the idea of a field was due to Michael Faraday, who performed what in retrospect was the next obvious experiment.

Eleven years later, Faraday turned Oersted's experiment inside out, reasoning that if an electric current could affect a magnet, possibly a magnet could be made to affect an electric current. He succeeded perhaps beyond his wildest dreams, demonstrating that if a magnet were moved through a coil of wire in such a way that the motion of the magnet continually changed, an electric current would flow in the wire. This is known as the principle of electromagnetic induction, and is the basis for the generation of electricity. I can't help but wonder what was responsible for the lag between Oersted's work and Faraday's; perhaps the fact that changing motion of the magnet—not constant motion—was required to generate the electric current, and failure to realize this was responsible for the aforementioned eleven-year gap.

The vast increase in the wealth of our society that has been made possible by cheap and widely available electric power has occurred because Faraday's discovery has enabled us to tap into the power of both the gravitational force and the Sun. The Sun's heat evaporates water from the ocean. It rises, cools, and falls as rain or snow at high altitudes. Gravitational force causes water to run downhill, which we can harness by placing dynamos inside of dams. Rotating the dynamos induces an electrical current that can be transported efficiently far from the source by means of cables. When we plug a device using an electric motor into an outlet, the electric current causes magnets to move, and it is this motion which enables the appliance to operate. The Sun's heat eventually evaporates the water that powered the dynamos, and the cycle starts again. And, of course, we rely on steam heated by burning coal, oil, and natural gas, and by nuclear fission, to turn dynamos, and produce electricity, as well.

Faraday also possessed a keen intuition regarding the electrical and magnetic forces he was studying. Many great scientific advances are made possible by new ways of conceptualizing phenomena. Faraday visualized electricity and magnetism as consisting of lines of force permeating space, with stronger forces creating a greater concentration of lines in a particular region. This method of visualizing electricity and magnetism led to the idea of a field, a type of mathematical description that occupies a central position in physics. Beyond that little taste, though, field theory will remain beyond the scope of this book.

The Relative Strength of Electricity and Gravity

I've done a lot of reading in science—after all, I'm a science junkie—and I have seen several different numbers quoted to describe the relative strength of the forces of electricity and gravity. Even though so distinguished a scientist as Martin Rees, who has received more prizes and awards than I have received handshakes, votes for 10^{36} as a measure of how much stronger the electric force is than the gravitational force, I'm going to vote for 10^{39}: a number I have seen in more than one place—and that makes the most sense to me. Let me see if I can convince you.

The most natural comparisons are the ones in which items are close to each other in kind. Let's take a look at the annual dog show of the Westminster Kennel Club to illustrate the idea. Every year they publish a list of Best in Breeds, and they also name a top dog—the Best in Show. While I am comfortable with the selection of the Best Beagle or the Best Golden Retriever, I really don't see how one can fairly compare beagles with golden retrievers. "Like to like" comparisons seem much more reliable.

So what does this have to do with comparing electricity and gravity? Gravity is *always* an attractive force, but electricity can be either a repulsive or an attractive force depending upon whether the charges involved are like or unlike. It seems to me that comparing the strength of gravity with a repulsive electric force is—well—a little repulsive (sorry about that). After all, I work out at the gym (occasionally), and I know that the same muscular configurations—the way the muscles are arranged in the body—have different strengths depending upon whether they're pulling or pushing. So a fair comparison would require an attractive electric force, which can be arranged by using a single electron (negative electric charge) and a single proton (positive electric charge).

The k in Coulomb's Law, $F = kqQ / r^2$, functions much as the big G we saw in Chapter 1. And, just as extraordinary efforts have been expended to determine the value of G, so have extraordinary efforts been expended to determine the value of k. Nevertheless, for the purposes of the computations that will be made in this chapter, we'll just say that $k = 9 \times 10^9$ newton-meters^2 / coulomb^2.

It doesn't matter how far apart we position the charges, because the factor of r^2 shows up in the denominator of both, and will therefore cancel out when we divide the electric force by the gravitational force.

For simplicity, we'll imagine the proton and the electron are one meter apart.

One coulomb is about 6.24×10^{18} electrons, so each electron carries a (negative) electric charge of $1 / (6.24 \times 10^{18}) = 1.6 \times 10^{-19}$ coulombs. The proton has an equal positive charge, and so the electric force between the two is $kqQ / r^2 = 9 \times 10^9 \times (1.6 \times 10^{-19}) \times (1.6 \times 10^{-19}) / 1^2 = 2.3 \times 10^{-28}$ newtons. Small though this may be, the gravitational force between the proton (with a mass of 1.67×10^{-27} kilograms) and the electron (weighing in at an even more diminutive 9.11×10^{-31} kilograms) is far less; as it is equal to $GmM / r^2 = 6.67 \times (10^{-11} \times 1.67 \times 10^{-27}) \times (9.11 \times 10^{-31}) = 1.01 \times 10^{-67}$ newtons. The ratio of the electric to the gravitational force between the two is therefore $2.3 \times 10^{-28} / 1.01 \times 10^{-67}$—about 2.3×10^{39}.

Of course, it's important to bear in mind that comparing different kinds of particles will give us different ratios. If one decides to compare the repulsive force, do you use the electron-electron repulsion or the proton-proton repulsion? The *electric* repulsive force is the same between two electrons or two protons, but because a proton weighs about 1,800 times as much as an electron, the gravitational force between two protons is more than three million times stronger than the gravitational force between two electrons.

That's not the only reason for my "like to like" comparison, however—it's also that the attraction between positively and negatively charged particles is hugely important to everyday life. Much of everyday life is chemical reactions, and chemical reactions are the result of electrons pirouetting happily from one atomic dance partner to another as the atoms involved in the chemical reactions change the partners with whom they are joined. If the ratio of electrical strength to gravitational strength were somewhat less, all the biochemical processes that we take for granted would be correspondingly more difficult. Walking, which represents the temporary defeat of gravitational attraction by biochemical processes—which are powered indirectly through the electric force—would be more of an effort. That doesn't mean that we couldn't do it, but our musculature would obviously have to be a lot stronger. The implications for us even developing such musculature are unclear, because we have no model for how evolution would work under such conditions—or if it would even work at all. Similarly, if the ratio of electrical strength to gravitational strength were somewhat more, that uncomfortable shock you get on a cold winter day when you touch a metallic doorknob might be more than just uncomfortable, it might be life-threatening.

Or it might not; evolution might have produced a mechanism to cope with this. There's simply no way for us to tell, because although we can devise scientific experiments to show how organisms cope with stronger or weaker electric or magnetic fields, all that coping is done in an environment in which the electric force is 10^{39} times as strong as the gravitational force, and we can't change that. We can, however, be thankful that that's the way it is—because here we are.

CHAPTER 7

THE
BOLTZMANN
CONSTANT

A lthough I can remember many of the events of my childhood connected with different aspects of science, there weren't many involving heat. I do remember brushing up against an electric iron on an ironing board, getting a small triangular burn (which my mother treated with butter, ice not being the cure-all in the 1940s that it would later become), and learning to be a little more careful when in the general vicinity of very hot objects. I also remember wondering why a warm bath was more comfortable than a humid day. I took the thermometer my parents used to take my temperature and stuck it into a bathtub filled with warm water, and I was somewhat surprised that my body temperature could get higher when I was really sick than the temperature that showed in the bath. My father, a generally knowledgeable man, explained to me that a hot humid day was uncomfortable because the body naturally lowers its temperature by perspiring, and it was harder to perspire on a humid day. That made sense. And, assuming the temperature in the water is cooler than about 310 kelvins (roughly your body temperature), you'll be losing heat to the water as you sit in it, and you do perspire (from those portions of your body not immersed in the water). There are questions that may be beyond the ability of science to answer, such as why men prefer showers and women prefer baths. But if there's one thing we know about, it's heat.

The Phlogiston and Caloric Theories of Heat

Like light, the nature of heat has been a major preoccupation of science for a long time. The Greeks, of course, thought of fire as one of the basic elements, and because fire seemed indistinguishable from heat, heat, too, was thought to be a substance. This remained a serious point of view for centuries, and the first post-Newtonian theory to incorporate this idea was the phlogiston theory, originated by Johann Becher and further developed by Georg Stahl in the late seventeenth century. It held that combustible substances contained phlogiston, a colorless, weightless substance that was liberated on burning, after which the substance was said to be dephlogisticated. Substances that burned easily were rich in phlogiston.

It's easy to critique an erroneous theory in retrospect, but the phlogiston theory explained—to some extent—both combustion and rust, in which iron appeared to be acquiring something. When a substance burned in an enclosed space, such as a bell jar, the burning soon ceased, and this was taken as evidence that the air had absorbed the maximum amount of phlogiston it could take up, like a sponge soaking up only a certain amount of water. It was also noted that when combustion took place in an enclosed space, living things could no longer breathe, so the phlogiston theory also seemed to partially explain respiration, as air that had absorbed too much phlogiston from a body made it impossible to breathe.

We now know that substances burn because of the presence of oxygen, a fact that was conclusively demonstrated by the experiments of the French chemist Antoine Lavoisier a century later. The discovery was the result of a busy century for chemistry. Gases as elements had been isolated and studied, and measuring equipment enabled gases to be weighed with accuracy, so Lavoisier was able to show that the weight loss of the oxygen in the air was exactly the amount of weight that the burned substances gained—and that combustion could not take place if oxygen were not present. Exit the phlogiston theory—and enter the caloric theory.

Lavoisier's caloric theory held that a fluid called caloric was actually the substance of heat, that the universe contained a constant amount of caloric, and that it always flowed from hotter to colder bodies. According to Lavoisier, a hot cup of coffee cools to lukewarm because the excess caloric in the cup of coffee flowed out to heat the area, and that a gas expanded when heated because the absorption of caloric made it larger. The theory wasn't right, but it did a pretty good job of enabling computations to be made involving the final temperatures of mixtures; at any rate, the caloric theory was considerably superior to the phlogiston theory, and there are elements of it that we retain today.

A couple of simple mixture problems will indicate the value of the caloric theory. Let's suppose that for each degree centigrade a gram of water is raised, it gains 1 unit of caloric, and for each degree centigrade that a gram of water is lowered, it loses 1 unit of caloric. If we mix 100 grams of water at 40 degrees centigrade with 50 grams of water at 10 degrees, what will be the temperature T of the mixture? The 100 grams of water at 40 degrees will lose $100 \times (40 - T)$ units of caloric, and the 50 grams of water at 10 degrees will gain $50 \times (T - 10)$ units of caloric. Since the caloric lost by the hot water equals the caloric gained by the cold water, we have

$$100 \times (40 - T) = 50 \times (T - 10)$$
$$4{,}000 - 100T = 50T - 500$$
$$4{,}500 = 150T$$
$$30 = T$$

So the final temperature of the mixture is 30 degrees. Had we used 100 grams of aluminum at 40 degrees rather than 100 grams of water in the mixture, we would have found by experiment that the temperature of the mixture was 19 degrees. Obviously aluminum has a different rate at which it gains or loses caloric, so let's assume that a gram of aluminum gains or loses c units of caloric for each degree centigrade it is heated or cooled. The 50 grams of water would gain $50 \times (19 - 10) = 450$ units of caloric, while the 100 grams of aluminum would lose $c \times 100 \times (40 - 19) = 2{,}100c$ units of caloric. Again, the amount of caloric lost by the aluminum equals the amount of caloric gained by the water, so $2{,}100c = 450$, and $c = 450 / 2{,}100 = 0.214$.

Caloric, over time, became calorie, which is the metric unit for energy, and which, although replaced by the joule in the modern system of measurements, still reigns supreme on the labels of food items everywhere. One calorie is the amount of heat required to a raise a single gram of water 1 degree centigrade. The mini-muffin I ate before writing this section contains 80 "large" calories—a large calorie being equal to 1,000 "small" calories, so the mini-muffin contains enough energy to heat a kilogram of water from 10 degrees centigrade (50 degrees Fahrenheit) to 90 degrees centigrade (194 degrees Fahrenheit). That number 0.214 that we obtained for aluminum is actually a parameter called "specific heat"; the specific heat of water is 1.000 and of aluminum 0.215 (actually, the final temperature of the aluminum-water mixture would have been 19.02 degrees, but I rounded things off a little in order to use simpler numbers in the computation). This represents the fact that it requires only 0.215 calories to heat 1 gram of aluminum 1 degree centigrade—and also explains why if you boil water in an aluminum pot on the stove, you can burn your fingers by touching the aluminum long before the water comes to a boil.

Forces, Work, and Energy

Forces change things. One's first take on forces is that they change positions of things; the gravitational force of the Earth pulls the apple down from the tree, and the electric force attracts charges of opposite sign and pushes away charges of like sign. However, Newton's law of inertia looks a little deeper than that: it says a body in motion continues at a constant velocity unless acted upon by a force. What a force does is change the momentum of an object. Momentum is the product of mass times velocity, and because mass doesn't change, except in those instances when Einstein's theory of relativity is relevant, the net effect of a force is to change an object's velocity. A change in velocity is what we call acceleration.

You need to exert a force to lift a 3 kilogram mass (which, on Earth's surface, weighs about 29 newtons)[1] 2 meters off the ground. Physicists quantify the amount of effort you have expended in doing this by describing your accomplishment as $2 \times 29 = 58$ newton-meters of work. If a constant force F is exerted through a distance d, the total work W done is given by $W = F \times d$. Applying calculus enables us to take this basic formula and generalize it to situations when the force is not constant and the path is not necessarily straight—such as the work done in carrying a leaking bucket of water up a spiral staircase.

So where does energy enter the picture? In fact, what *exactly* is energy? The word *energy* is derived from the Greek word *energeia*, meaning "active" or "working." The first person to quantify energy seems to have been Gottfried Leibniz, a contemporary of Newton and a competitor for the invention of calculus (they actually seemed to have come up with a lot of the same ideas at roughly the same time, but Newton is generally accorded the honor).

It's not hard to see how energy could have entered the picture. Suppose that an object with mass m is moving with constant acceleration a for a period of time T during which its velocity changes from an initial value v to a final value V. The force exerted on the object is ma; let's assume that the force moves the object through a distance D so that the total work performed on the object is maD. Since the acceleration is constant, $V = v + aT$, so $a = (V - v) / T$. Since the velocity increases at a constant rate, the distance traveled during the time T can be computed as the average velocity $(V + v) / 2$ times T. Therefore,

$$
\begin{aligned}
W &= maD \\
&= m \times (V - v) / T \times ((V + v) / 2)T \\
&= \tfrac{1}{2}m\,(V - v)\,(V + v) \\
&= \tfrac{1}{2}mV^2 - \tfrac{1}{2}mv^2
\end{aligned}
$$

Therefore, the work done is given by the difference between $\tfrac{1}{2}mV^2$ and $\tfrac{1}{2}mv^2$; it is as if we've exchanged the difference in these quantities for the work done, just as if we buy a cup of coffee at Starbucks, we've exchanged the difference between our after-purchase wealth and our before-purchase wealth for the cup of coffee.

The quantity $\frac{1}{2}mv^2$ is what is known as kinetic energy; it only occurs when an object is moving, because $v = 0$ if it's not moving. Because velocity is relative to a particular reference frame, kinetic energy is as well. If you are sitting still in a car moving 60 miles per hour, you have no kinetic energy relative to the moving car but a whole lot of kinetic energy relative to the road, and so if the car crashes, your kinetic energy has to go somewhere—and it's a lot better that it be transferred to a seatbelt or an airbag than to the dashboard or the windshield of the car.

There's another kind of energy that isn't recognized in the calculation above. Albert Einstein gives a good example of this in his book *The Evolution of Physics*.[2] Imagine a roller coaster poised motionless at the highest point of the ride. It has no kinetic energy at all—because it has no velocity relative to the roller-coaster track—but it has a substantial amount of potential energy that is available from letting the Earth's gravitational field do its work. And, of course, the Earth's gravitational field will do just that, and at the bottom of the track some of the energy from the Earth's gravitation field will have been exchanged for kinetic energy as the roller coaster reaches its highest velocity. Then up you go, losing kinetic energy from the moving roller coaster as its velocity decreases, but exchanging it for increased potential energy from the Earth's gravitational field as you move further away from the Earth, the source of that potential energy. The ride continues, exchanging one form of energy for another, back and forth.

But the ride eventually comes to an end. It might initially seem that it can go on forever—and the few roller-coaster rides that I have been talked into, usually by girlfriends in front of whom I did not want to appear to be a chicken, *seemed* to go on forever. However, there is another horse in the energy derby beside kinetic energy and potential energy— friction. The movement of the car on the track heats the track by friction, and late in the eighteenth century, this insight opened the way for the modern theory of thermodynamics.

The First Law of Thermodynamics

The death knell for the "heat is a substance" theory can be found in a paper by Sir Benjamin Thompson, better known as Count Rumford. (I have no idea why scientists with titles are better known by their titles than by their names; I guess it's a Euro thing.) Rumford, a British loyalist from New Hampshire, fled America after the Treaty of Paris for Munich, where he investigated the manufacture of cannon. At the time, a cannon was made by drilling a cylindrical hole in a steel cylinder. The scraps of metal tossed out by the drill were extremely hot—Rumford actually made detailed measurements of the temperature of these scraps. It occurred to Rumford that if heat were a substance, as the caloric theory maintained, it could not be inexhaustibly generated as seemed possible from the mechanical drilling of cannon. He published his observations and conjectures in 1798 in his most important scientific paper, *An Experimental Enquiry Concerning the Source of the Heat Which Is Excited by Friction.*[3] Just in case the reader of *An Experimental Enquiry* was still clinging to the notion that heat was a substance, Rumford delivered the knockout punch. "It is hardly necessary to add," he wrote, "that anything which any *insulated* body, or system of bodies, can continue to furnish *without limitation*, cannot possibly be a *material substance*; and it appears to me to be extremely difficult, if not quite impossible, to form any distinct idea of anything, capable of being excited and communicated, in the manner the Heat was excited and communicated in these Experiments, except it be *Motion*."[4]

Rumford was right—but he didn't realize in exactly how many different ways he was right, and what the consequences were. If the seventeenth century was characterized by the birth of mechanics and the eighteenth century by the development of chemistry, the nineteenth century can be seen as the century in which energy in many of its forms came to be understood and utilized.

James Joule was at the forefront of this revolution. Born to a reasonably affluent brewer, he and his brother were tutored by no less a scientist than John Dalton. Granted, many great scientists and mathematicians have done some tutoring to augment their income, but this seems to be roughly on the same level as Alexander the Great hiring Aristotle to teach him geometry. Dalton suffered a stroke after two years of this employment, but his emphasis on science and method made a lasting impression on Joule.

Joule's greatest contribution was his realization that different forms of energy were equivalent in the same sense that different forms of money are equivalent. Dollars and euros are different forms of money—they look different, but the bottom line is that there is an exchange rate between the two that enables us to compare the prices when they are quoted in one currency or the other. Joule's observation is very clearly stated in his classic 1845 paper, "On the Mechanical Equivalent of Heat,"[5] where Joule remarks that "the mechanical power exerted in turning a magneto-electric machine is converted into the heat evolved by the passage of the currents of induction through its coils; and, on the other hand . . . the motive power of the electro-magnetic engine is obtained at the expense of the heat due to the chemical reactions of the battery by which it is worked."[6]

Joule constructed a very simple apparatus, in which a falling weight turned a paddle immersed in a container of water. The mechanical energy derived from the falling weight was converted into heat by stirring the water, and Joule performed a number of measurements to discover the rate of exchange. It is difficult to imagine scientific discourse taking place nowadays between an academic and a brewer, but the twain met much more freely in nineteenth-century Britain, and one of the attendees at a talk given by Joule in 1847 to the British Association in Oxford was William Thomson, the recently appointed Professor of Natural Philosophy at the University of Glasgow. Thomson, later to become Lord Kelvin, was intrigued by Joule's results. Joule married later that year, and while he and his bride were honeymooning in Chamonix, they accidentally ran into Thomson. Her husband, clearly a romantic, interrupted the honeymoon so he and Thomson could attempt to measure the temperature difference between the top and bottom of the 1,200-foot waterfall. I could find no record of Amanda's reaction—but then, I could find no record of this having any adverse effect on the marriage, either.

Sadly, Joule's wife and daughter died barely five years later. Joule did further work with Thomson—in fact, he and Thomson discovered that when a gas is allowed to expand without doing work, its temperature decreases, and this eventually led to refrigeration. Unfortunately, his passion for science also resulted in the demise of the family brewery; fortunately, he managed to obtain a British civil-service pension—not having a family, he was able to live comfortably until his death. The work for which he is best known is commemorated at the top of his tombstone. Mechanical work, as has been discussed earlier in this chapter, can be measured in newton-meters, but nineteenth-century British physicists preferred using foot-pounds. Joule's work showed how the universe converts mechanical work into heat energy, which in those days was measured in British thermal units (BTU), defined as the amount of heat needed to raise 1 pound of water 1 Fahrenheit degree at atmospheric pressure. The number on Joule's headstone is 772.55, and it commemorates an experiment he did in 1878, when he showed that it took 772.55 foot-pounds of work to generate 1 BTU.

Though 772.55 foot-pounds per degree Fahrenheit is sometimes known as Joule's constant, it is not a truly fundamental number, in the sense of this book, as it's really just another form of water's specific heat. All substances have them, and although water is an exceptionally important molecule on Earth, you can be sure that, should a species evolve that relies on methane as we do on water, it would be methane's specific heat that filled the place of Joule's constant.

Joule's constant may not be truly fundamental, but it did lead directly to the first law of thermodynamics, which states that energy can be transformed from one form into another—such as mechanical energy being transformed into heat, or kinetic energy into potential energy—but can be neither created nor destroyed. If we take another look at Einstein's roller coaster and start with the coaster at its highest position, its total energy is all potential energy. Its first descent to the bottom generates a certain amount of heat through friction against the track. At the bottom, the total energy of the system is the sum of its kinetic energy, the heat energy produced by friction, and the lower potential energy because it is closer to the center of the Earth. When the coaster now ascends, it again creates more heat through friction, and as a result it cannot ascend to its initial height, for to do so would require the same total energy that the system originally possessed before the ride started—and some of that energy has been dissipated as heat.

The Second Law of Thermodynamics

The various forms of energy are interchangeable, but it still seems that the universe prefers heat energy. It is relatively easy to transform mechanical energy into heat—in fact, this often happens both unbidden and unwanted. When mechanical energy is transformed into heat, it is often energy wasted in the sense that we can't do anything with it. Going back to Einstein's roller coaster, the friction warms the track, but what good is that? This isn't to say it's impossible to do the reverse: ever since the invention of the first steam engine, we have been transforming heat into mechanical energy. This technique still powers much of our transportation via the internal combustion engine in automobiles and the jet engine in airplanes, but it does so inefficiently, in the sense that much of the heat is wasted. In the internal combustion engine, the heated gas in a cylinder expands; this expansion moves a piston that is connected to the driveshaft of the car, and through an ingenious system of rods and cams, the back-and-forth motion of the piston is transformed into the rotary motion of the wheels. However, at the same time that the gas is being heated to drive the piston back, the sides of the cylinder are being heated as well—to no one's benefit. Too much heat will warp the cylinder walls, so the cylinder is lubricated to reduce the friction. Nonetheless, the excess heat is going somewhere other than the expansion of the gas in the cylinder, and so your car has an intricate cooling system to prevent the excess heat from doing damage.

The question of how efficient a heat engine can be was first studied by Sadi Carnot, a French physicist and military engineer. A puzzling but commonplace observation about heat energy is that heat engines work by heating something; if everything is at the same temperature, there is no way to extract the heat. In Joule's experiment where the falling weight heats the water, once the water is heated, there's no way to use that heat unless it can flow to something cooler. The universe transforms kinetic energy to potential energy back and forth with considerable efficiency—as a planet approaches the Sun, the potential energy it possessed is transformed into kinetic energy as it moves more rapidly in accordance with Kepler's laws; as it swings away from the Sun, that kinetic energy is transformed into potential energy. This transference is highly efficient, as the planets maintain approximately the same orbits for millions of years.

On the other hand, when hot water is mixed with ice cubes, the end result is a glass of lukewarm water—and no one has ever witnessed ice cubes spontaneously appearing in a glass of lukewarm water at the same time that the water that isn't frozen into the ice cubes heats up. This is a fundamental property of heat. Carnot's magnum opus, *Reflections on the Motive Power of Fire*,[7] expresses this succinctly. "The production of motive power," writes Carnot, "is therefore due in steam engines not to actual consumption of caloric but to its transportation from a warm body to a cold body."[8] Even though Carnot used the caloric theory to describe his results and observations, his ideas are independent of it—whether heat is a substance such as caloric or a form of energy, motive power can only be produced by the transportation of heat from a warm body to a cold body.

The laws of thermodynamics initially arose out of observation and experimentation, unlike the law of conservation of energy in mechanics, which is a mathematical deduction from Newton's laws of motion. Carnot's observation, cited in the previous paragraph, was probably so well-known that no one really bothered to consider its significance; things seemed to spontaneously cool down, but it required the motive power of fire to get things to heat up. The first formal codification of the second law of thermodynamics is due to the German physicist Rudolf Clausius, who said that heat cannot flow spontaneously from a colder substance to a higher one—not exactly the way Carnot phrased it, but close. It might actually have occurred to Carnot to phrase it this way—but we'll never know, as he died of cholera at age thirty-six, and many of his books and writings were burned with him in an attempt to prevent the disease from spreading.

Several equivalent formulations of the second law of thermodynamics followed. The best known one is due to Lord Kelvin, who put it in terms of processes in which heat is converted to work (which, after all, is what a heat engine does). His formulation of the second law is that a heat engine cannot work with perfect efficiency, converting all the heat into work. A colloquial way to phrase the first law is that you can't win; there's no way to get free energy from the universe. Kelvin's statement of the second law could be colloquially phrased as you can't break even; the universe, like some gigantic Thermal Revenue Service, charges a heat tax when you try to extract work from heat.

Possibly the most intriguing formulation of the second law comes from a mathematical function Clausius invented, which he called entropy. I work with some very bright young children, and recently one asked me what entropy was. This sent me scurrying to look for a good intuitive definition of entropy, and I found one I really liked—entropy is a measure of the amount of unusable energy in a system. If we were to look at the example of the glass with ice cubes and hot water becoming a glass with lukewarm water, there is obviously more unusable energy in the glass of lukewarm water, so its entropy is greater than the entropy in the glass containing hot water and ice cubes. Clausius defined the entropy of a system as the sum of the quantities $\Delta Q / T$, where T is the temperature of an item in the heat-bookkeeping ledger for the system, and ΔQ is the amount of heat gained or lost by that item—positive if the item gains the heat, negative if it loses it.

For example, imagine a system with two objects in it, one at 100 degrees, the other at 200 degrees. Suppose that we compute its entropy, and then allow a single heat transaction; the object at 200 degrees gives up 1 calorie to the object at 100 degrees. The entropy has changed by $^+\frac{1}{100}$ (as the object at 100 degrees absorbs the calorie) $- \frac{1}{200}$ (as the object at 200 degrees loses that calorie). The net change is $^+\frac{1}{100} - \frac{1}{200} = \frac{1}{200} > 0$, so the entropy of the entire system has increased. Because heat always flows from hotter items to colder ones, this single transaction typifies what always happens; entropy always increases.

This led to a concept called "the heat death of the universe." What happens when all the possible heat transactions have been made, and everything is at the same temperature? The answer given by thermodynamics is simple—nothing, and that's all she wrote. Entropy is as high as it can possibly go, and there's no way for it to increase. The heat death—actually, it's a *really* cold death—refers to the fact that no additional work can take place when entropy has reached its maximum.

One big question remained to be answered—just why does heat flow from hotter to colder substances?

Ludwig Boltzmann

The answer to the question of why heat flows from hotter to colder substances came from one of the most appealing scientists I could have wished to encounter. One of the pleasures of writing a book such as this is that you get to do some in-depth reading about people as well as discoveries. In reading about Ludwig Boltzmann, I was struck by how truly complete an individual he was, not just as a scientist, but as a human being.

I'm a sucker for anyone with a good sense of humor. Boltzmann's was legendary—at least among his circle of friends and acquaintances. It was a gentle and self-deprecating sense of humor that is only rarely seen these days. He described a sumptuous dinner of which he partook by saying, "In the restaurant of the Northwestern station I consumed roast young pig with sauerkraut and potatoes and drank a few glasses of beer in great contentment. My memory for numbers, otherwise reasonably reliable, always retains the number of beer glasses rather poorly."[9] By all accounts, he had a happy marriage, and cherished his three daughters, for whom he held dances (remember, this was the nineteenth century) at which he played piano. He was frequently the life of the party. A colleague described him in the following fashion, "Because of his capacity for communication, his ready wit and his clever and funny ideas, he soon became the center of every gathering of ladies and gentlemen, and dominated the conversations."[10]

It is rare that a top-flight scientist is also a top-flight teacher. I have been fortunate to know a number of top-flight mathematicians, and although a few were terrific teachers, more than a few were absolutely terrible. Boltzmann, however, was in the top tier in both categories. One of his students was Lise Meitner, who was later to play a central role in one of the epochal experiments that define the twentieth century: the discovery of nuclear fission. Meitner attended Boltzmann's lectures for four years, from 1902 through 1905, and wrote, "His lectures were the most beautiful and stimulating that I have ever heard. . . . He was himself so enthusiastic about all he was teaching that we left every lecture with the feeling that an entirely new and wonderful world was being opened to us."[11] As a teacher, I cannot imagine being paid a higher compliment.

Before I started writing this book, I only knew Boltzmann from his discoveries—and from his demise. Boltzmann suffered from depression, hard though that may be to imagine from the description already presented, and while on vacation near Trieste, he hanged himself. I first learned about Boltzmann, and about his death, around the time I was taking a poetry course in which I read the poem *Richard Cory*, by Edward Arlington Robinson. The fictional Richard Cory was a wealthy man in a poor town; the poem is a first-person narrative by one of the poor townsmen. The last verse has stuck with me throughout my life, perhaps because of the juxtaposition of studying the works of Boltzmann and reading about his suicide.

> *So on we worked, and waited for the light,*
> *And went without the meat, and cursed the bread.*
> *And Richard Cory, one calm summer night,*
> *Went home and put a bullet through his head.*[12]

How incredibly sad that such brilliant individuals as Wallace Carothers and Ludwig Boltzmann, to say nothing of the myriads of others cursed with depression, should have their judgment and emotional stability so clouded that they see nothing else to do but to take their own life. And what a loss for those who care about them, and for the rest of us who could have benefitted from the discoveries they might have made.

Statistical Mechanics

Boltzmann was a great believer in the atomic theory. From the perspective of the twenty-first century, this sounds a little like being a great believer in the heliocentric theory originally espoused by Copernicus; that's the way things are—duh. During Boltzmann's era, however, the atomic theory—and so Boltzmann himself—was under continual attack from some very prestigious physicists; always finding himself on the defensive may have been one of the factors contributing to Boltzmann's depression.

If you believed, as Boltzmann did, in the validity of the atomic theory, and you knew, as Boltzmann did, how large Avogadro's number is, you would realize that the description of a liter of an ideal gas as a collection of roughly 10^{23} molecules—10^{23} individual objects—was completely impossible. Even tracing the course of a single molecule in this melee would be completely impossible. Boltzmann was one of the prime contributors to statistical mechanics, an aptly named discipline that applied the mathematics of statistics to the mechanics of how the gas molecule behaved.

One cannot describe the economic activity of the United States by trying to describe the economic activity of each individual. There's just too much data. However, we can form a distribution for all the important economic variables, such as what percentage of people have annual incomes between \$50,000 and \$75,000, etc. This is an incomplete description, but it suffices for many purposes. The same thing can be done for the molecules in a jar of gas, as for many practical purposes, knowing the distributions of the positions and velocities of the molecules, such as what percentage are moving between 30 and 35 centimeters per second, will suffice to describe the system. This is the approach of statistical mechanics, and in the latter portion of the nineteenth century, this approach recorded many triumphs. One of them was the discovery of Boltzmann's constant, which relates the energy of individual particles to the temperature of a bulk object.

Boltzmann's Constant

Boltzmann didn't have to look very hard for his constant, known as k; he simply divided the ideal gas constant R by Avogadro's number. Nevertheless, Boltzmann's constant k appears in several important equations in the kinetic theory of gases, statistical mechanics, and thermodynamics, and enables us to deal mathematically with the fact that although we can talk about the gross characteristics of a system—such as the temperature of a tank of gas—not all the molecules in it have the same energy. The molecules in a tank of gas, for example, randomly bump into one another, and these collisions cause some of them to be moving more rapidly than the average speed, and some more slowly. However, the average translational energy possessed by a monatomic molecule in an ideal gas can be shown to be $\frac{3}{2} kT$, where T is the temperature in degrees kelvin. Since the translational energy of the molecule with mass m and velocity v is $\frac{1}{2} mv^2$, assuming that no energy is lost to rotation we have $\frac{3}{2}kT = \frac{1}{2}mv^2$. So the temperature increases as the square of velocity of the molecule.

Solving the equation for T, we have $T = \frac{1}{3}mv^2 / k$. Absolute zero is the special case when $v = 0$, but setting $v = c$ gives us something perhaps even more interesting—the highest temperature any particular substance can reach. Let's consider radon, the heaviest noble gas. Noble gases are about as close to ideal as the real world can get, they're called noble because they rarely interact with the other elements, much as members of the nobility find it difficult to interact with the hoi polloi. A mole of radon has a mass of about 222 grams, so a single atom of radon weighs $222 / (6 \times 10^{23}) = 3.7 \times 10^{-25}$ kilograms. The speed of light is about 3×10^8 meters per second, so if it were possible to get an atom of radon to travel very close to the speed of light, its temperature would be $\frac{1}{3} \times (3.7 \times 10^{-25}) \times (3 \times 10^8)^2 / (1.38 \times 10^{-23})$ kelvins, or about 2.7×10^{14} degrees kelvin, or 4.8×10^{14} degrees Fahrenheit. Toasty. However, those of us who have lived less-than-exemplary lives can take some consolation in that hell cannot possibly be as hot as this, as I have yet to hear any theologian espouse the position that hell consists of radon atoms rushing around at close to the speed of light.

Returning to the collection of molecules in an ideal gas, Boltzmann was able to show that the fraction of molecules in an ideal gas with energy E was proportional to $e^{-E/kT}$. Different values of T give different curves, but—to run the risk of oversimplifying a bit—they all look roughly like the bell-shaped curve. For low values of T, the curves have a very narrow central peak, as T increases the curve becomes lower and fatter. This isn't so surprising; with higher temperatures the molecules are moving faster, and if you look at a road where the speed limit is 30 miles per hour, a lot of cars will be moving really close to 30 miles per hour, but when the speed limit is 70 miles per hour the little old ladies, who are generally not of the ilk described in the song "The Little Old Lady from Pasadena,"[13] can be found hovering in the general vicinity of 50 miles per hour, with the soccer moms at 60 and the teenagers and sports car drivers above the speed limit.

The *pièce de resistance* for Boltzmann's constant is the equation to be found on his tombstone. Boltzmann's constant is expressed in the same units as entropy, which you will recall is formally the sum of heat divided by temperature. In order to understand Boltzmann's equation, let's consider a really small sample of an ideal gas with two molecules, A and B. Let's suppose that A is moving at 50 cm/sec and B is moving at 100 cm/sec. We measure the temperature of the gas and find it has a certain value, which we denote by T. If the molecule A had been moving at 100 cm/sec and B at 50 cm/sec and we measured the temperature, we would have gotten the same result T. The value of T is called a *macrostate* of the system, the two arrangements of the speeds of A and B (50 cm/sec and 100 cm/sec) are called *microstates* corresponding to the macrostate T. Obviously, when you have a mole or so of an ideal gas, there are a lot of microstates corresponding to the same macrostate. The higher the temperature, the more microstates there are corresponding to the same macrostate. Again, a gas is like cars on the road; there are a lot more ways you can get 100 cars to travel with an average speed of 50 miles per hour than you can with an average speed of 10 miles per hour.

This relationship is expressed by the equation $S = k \ln W$, where S is entropy, k is Boltzmann's constant, and W is the number of microstates corresponding to the macrostate that resulted in the entropy S. The microstate-macrostate picture also finally explains why heat flows from higher temperatures to lower temperatures. It's simply a matter of probability. Any one microstate is just as likely as any other microstate, but the macrostate described by the glass of warm water has many more microstates associated with it than the macrostate described by the ice cubes and hot water configuration. Interestingly enough, this also opens the door for the glass of warm water to morph into the glass with ice cubes and hot water—it's just so tremendously unlikely that we haven't seen it yet, and it doesn't figure to happen in the entire lifetime of the universe.

As I noted above, Boltzmann's constant is the quotient of two other fundamental constants—both of which we have already encountered. This might seem cheap, somehow, but it is not the only big idea in science for which what looks like a simple reshuffling of previously known ingredients is actually profound enough that it gets named for the individual who performed the reshuffle. I first encountered this when I saw that the French mathematician and scientist d'Alembert had taken the equation $F = ma$ from Newton's second law, subtracted ma from both sides to obtain $F - ma = 0$, and this had been renamed d'Alembert's principle of least virtual work. I was simultaneously appalled and intrigued—possibly here was a potential route to scientific immortality. Just divide the gravitational constant G by the speed of light c, and there it was—Stein's constant. Visions of this being inscribed on my tombstone were quickly squelched when I realized it wasn't enough to perform the reshuffling, one actually had to show the reshuffling accomplished something significant. So, for my tombstone, I've decided to go with "This was the last item on the list."

CHAPTER 8

THE
PLANCK
CONSTANT

I t might not be surprising to learn, given what we know about what Max Planck would eventually accomplish, that Planck entered university at the age of sixteen. I rarely see sixteen-year-olds in my classes, but when I do, I can generally count on them to be star performers— or even have a decent shot at greatness. Certainly Planck was a good student: in his last three years at the Maximiliansgymnasium in Munich, he ranked eighth in a class of twenty-three, third in a class of twenty-one, fourth in a class of nineteen.[1] He wasn't, however, a star. Yet he was a beloved student, "deservedly, the favorite of his teachers and classmates."[2] I can't speak to what made him a favorite with his classmates, but based on my own experience, I can hazard a good guess as to what made him a favorite with his teachers. Sure, every teacher would love to have a truly brilliant student, but it is even more gratifying to have a good student who gives the proverbial 110 percent. The question that remains is how he, a good-not-great student, managed to turn physics on its head—especially considering what happened next.

Planck was interested in physics, so he sought out the advice of Philipp von Jolly of the University of Munich, which he was to enter. Jolly, primarily an experimental physicist, was not particularly sanguine about the future of physics as an intellectual discipline. He told

Planck that "in this field, almost everything is already discovered, and all that remains is to fill a few unimportant holes."[3] (I can't *ever* imagine telling this to a student or prospective student, possibly because mathematics deals not only with so many different subjects but also because infinity is a very important part of mathematics, although there have been periods—such as the present—when the problem was a lack of jobs and an excess of applicants. Fortunately, I arrived on the scene when the opposite was true.) Planck replied to Jolly that he didn't wish to discover new things, only to understand what was already known in the field.

Neither Jolly nor Planck could have been more wrong. It's more than a century later, and not only have the holes to which Jolly referred not been filled up, vast yawning caverns of ignorance remain to be explored and there are undoubtedly more undiscovered holes still remaining than in the annual global production of Swiss cheese. Although Planck may not have wished to discover new things, he indeed did so—and the new things that he discovered constitute the greatest revolution in physics since Newton first set pen to paper.

Planck made quick work of college—he embarked on his doctoral dissertation at the relatively tender age of twenty and finished it in four months—but his work still didn't impress his elders. His dissertation, on the second law of thermodynamics, made hardly any impact on the upper levels of the German physics establishment. Gustav Kirchhoff, the giant who had discovered spectroscopy and had made major contributions to the theory of electrical circuits, considered it to be wrong. Two other giants, Hermann von Helmholtz (who crystallized the idea of the conservation of energy) and Rudolf Clausius (who introduced the idea of entropy), didn't even bother to read it. Planck spent five years without an academic appointment, until his father, who had a good deal of clout at the University of Kiel, helped Planck get hired there in 1885 as the equivalent of an associate professor. Shortly thereafter, he won second prize in a competition given by the Philosophical Faculty of the University of Gottingen on the nature of energy. No first prize was awarded; the obvious implication is that Planck's essay was judged to be superior to the other entries but unexceptional. Planck, however, caught yet another break—even though Gottingen was unimpressed, Berlin was, and offered him a position, again as an associate professor—same rank, but a more prestigious university. And, ironically, the position he took over had been relinquished by the same Gustav Kirchhoff who felt that Planck's doctoral thesis was in error.

Vindication was to come several years later. Planck's thesis had by now become a recognized work of importance, and he had to loan it out so often that it was almost ready to fall apart (this was not an era in which one could simply copy and paste), and by 1892 Planck achieved the rank of full professor at Berlin. It was about this time that Planck became interested in the problem that was to vault him into the ranks of the immortals.

Heat and Light

If you own an electric stove, you've undoubtedly noticed that when you turn on one of the heating elements, the colors gradually change from a dull red to a bright orange. At least that's what happens on my stove; if yours is capable of generating a lot more heat than mine, you'll notice that the color will gradually change to a yellowish white, then a bluish white. Of course, if that happened, you'd have to have a pretty special stove. If the metal becomes yellowish white, it's in the range of 1,600 kelvin (2,400 degrees Fahrenheit). A standard heating element, made of iron, would be melting, and so would the top of the stove; and of course the manufacturers have taken safeguards to make sure this doesn't happen. Bad things can happen when the heating element on a stove runs for too long (I generally check several times to see that the stove has been turned off when I leave the apartment, one of the symptoms of advancing age), but not because the heating element has melted.

We know today how color and temperature relate, but the investigation of it was difficult. Kirchhoff, who was one of the first to take it on, was able to demonstrate something of fundamental importance: the color does not depend upon the material being heated or upon its configuration. Whether you have an iron spiral, as the heating elements on my stove are configured, or a tungsten wire filament, as Thomas Edison was to use in his first successful electric light, the color sequence depends only on the temperature.

Color is characterized by the wavelength of the electromagnetic wave an object is emanating; the longest visible waves are a dull red, but still longer waves are infrared, below red. We don't think of ourselves as radiant objects, but we are—our body temperature is generally in the range of 310 kelvin, and most of the heat energy that our bodies radiate is in the infrared, which is why we can be detected in a dark room by an infrared sensor. However, not all the energy that is radiated by a hot object is radiated at a specific color. The color that we see from a hot object is the wavelength where the preponderance of the radiant energy is concentrated, but in reality every object from liquid helium to the hottest star radiates its energy at different wavelengths.

Physicists approached the relationship between temperature and light by considering an object known as a blackbody in thermal equilibrium, meaning its temperature is stable. A blackbody is something that perfectly absorbs and emits electromagnetic radiation. The classic example of a blackbody is a cavity radiator: it's a hollow sphere with a very small opening; there is so little space for the radiation to emerge from the sphere that very little of it does—it bounces around inside, heating all portions of the interior equally. A major quest of nineteenth-century physics was to determine—theoretically if possible, empirically if necessary—the curves that represented the distribution of radiation that emerged at different wavelengths from a blackbody. For each temperature, physicists predicted, there should be a different curve.

These curves were of more than theoretical interest. Electric lights were clearly the wave of the future, and electric lights produced radiation via heating. Knowledge of these curves would allow engineers to design lights that produced light while wasting as little heat as possible. In fact, the Siemens family had donated money to found an institute in Berlin to integrate theoretical science with the practical needs of industry.[4] Wilhelm Wien, one of the institute's scientists, was able to make significant inroads into this problem.

To understand Wien's results, it is necessary to understand some basic terminology involving waves. I've always enjoyed living near large bodies of water, and I live in Southern California near the Pacific Ocean. There's something pleasurable about standing near the edge of the ocean—or, if the water is not too cold, actually going in—and watching the waves roll up and onto the beach. The waves can be characterized by how high they are—their amplitude—and how many crash onto the beach in a given time period—their frequency. Wave frequency is denoted by the Greek letter ν (pronounced "nu"), and the color of light is determined by its frequency in cycles per second, also known as hertz.

There is an incredibly large range to the frequency of electromagnetic waves. Radio and television waves have relatively low values, in the tens or hundreds of millions of cycles per second. At the upper end of the range are the powerful gamma rays usually produced by massive explosions; these rays have frequencies in excess of 10^{18} cycles per second. The portion of the electromagnetic spectrum that we can see is only a small fraction of the entire spectrum; the frequency of red light is on the order of 4×10^{14} cycles per second, and the frequency of blue light about 7.5×10^{14} cycles per second. Infrared light has a slightly lower frequency than that of red light; ultraviolet light has a slightly higher frequency than blue light.[5]

Wien performed his experiments in the blue portion of the spectrum. He discovered that if T were the temperature in degrees kelvin and ν the frequency of the light, the expression $I(\nu,T)$ for the intensity of the light being radiated at frequency ν by a blackbody heated to T degrees could be approximated by $I(\nu,T) = A\nu^3 e^{-B\nu/T}$, where A and B were positive constants whose values Wien could determine empirically. That was good enough for his industry sponsors; they would be satisfied simply by having numerical values that they could use, but theorists would never be satisfied until they knew why the values of A and B were what they were. So, for example, if a theorist were to see the number 186,000 (or thereabouts) in an expression, he would undoubtedly wonder why the speed of light was appearing in that expression, and would try to develop a theoretical derivation to explain it. Wien's formula was known as Wien's radiation law or Wien's approximation—it worked, but it didn't really advance the theory of radiation all that much.

The Rayleigh-Jeans Law and the Ultraviolet Catastrophe

At the same time Wien was working, across the North Sea, two men, Lord Rayleigh and James Jeans (later to be *Sir* James Jeans), were trying to predict the same intensity curves that Wien was determining empirically.

Their starting point was the concept of equipartition of energy: the idea that for a system in thermal equilibrium, such as a blackbody, the total energy available in the system is divided equally between all available forms of energy. A gas molecule, for example, may possess both translational energy from the speed at which it is moving, and rotational energy from the way it rotates. This idea had proved quite fruitful in statistical mechanics,[6] resulting in the Maxwell-Boltzmann velocity distribution for noble gases, and so it was certainly a reasonable assumption for Rayleigh and Jeans to make in their analysis of radiation. The result of their theoretical analysis was an expression for $I(v,T)$ that differed substantially from Wien's empirical description. The Rayleigh-Jeans Law, as it became known, stated that $I(v,T) = 2(k / c^2) v^2 T$, the constant k being Boltzmann's constant, and c the speed of light. This formula had a significant advantage over the Wien approximation in two respects. First, its constants were known physical constants, rather than empirically determined ones. Second, it fit the measured intensity curves for red light better than the Wien approximation.

It also had one major disadvantage: it was obviously wrong. The numbers k and c were fixed, and so if the temperature T of the blackbody were fixed as well, the intensity $I(v,T)$ increased as the square of the frequency v. This meant that, as the frequency became larger and larger, the intensity of the light at that frequency would increase without limit. This obviously didn't happen; all the actual intensity curves peaked at a certain frequency and then fell off at higher frequencies; a red-hot iron bar is red and not scorchingly blue and violet at the same time. The physicist Paul Ehrenfest gave this result an attractive name:

the ultraviolet catastrophe. (If I were enough of a rock musician to start a garage band, I would definitely call it the Ultraviolet Catastrophe.) The ultraviolet catastrophe referred to the fact that as the frequency of light increased toward the ultraviolet, the Rayleigh-Jeans expression not only failed to match the observed intensities but predicted obviously ridiculous ones as well.

This was the situation at the turn of the century. The stage was set for Max Planck to revolutionize physics.

Enter the Quantum

Ever since Kirchhoff had demonstrated that it didn't matter what substance the blackbody was made of, or what its shape was, theorists had been free to use any model they wanted. Planck chose to model the system as a collection of simple harmonic oscillators; a metallic spring is a good example of a simple harmonic oscillator. Molecules vibrate somewhat like springs, so it's not really all that far-fetched an assumption. Planck also started along the same lines that Rayleigh and Jeans did; he assumed that the energy emitted by the oscillators could come in any measurable quantity. He, too, ran into the ultraviolet catastrophe using this approach.

Then, one day, he made a different assumption—one that he told his son he felt was as revolutionary an idea as had ever occurred to Newton or Maxwell.[7] Instead of assuming that the oscillators could radiate energy at any level, he assumed that there was a number h so that if an oscillator was emitting energy at a frequency v, that energy had to be an integer multiple of hv—hv, $2hv$, $3hv$, and so on.

This assumption had an immediate consequence—it eliminated the catastrophe by putting an upper limit on the frequency of light any given blackbody could emit. Because the total energy of a blackbody has to be finite—let's call the amount E—the intensity of radiation must have an upper bound. If all the energy in the blackbody were put into a single oscillator (unlikely though that may be) radiating at a frequency v, the largest possible value for v would occur if the energy $E = hv$. If this were the case, $v = E / h$ would be the highest frequency that occurred; if that one oscillator was radiating energy at a multiple of hv, such as $2hv$, the frequency would max out at $E / 2h$; and if there were other oscillators radiating, that would decrease the value of E available to any given oscillator. That meant that the ultraviolet catastrophe could not occur, because one could not obtain arbitrarily high frequencies to create the catastrophe. Planck's assumption was the definition of ad hoc, but it did enable him to derive the following formula for intensity: $I(v,T) = (2hv^3 / c^2) / (eh^{v/kT} - 1)$. The formula had at least three attractive features. First, it doesn't run into the problem of having the intensities get arbitrarily large; Planck's equation multiplies a power function (the variable hv raised to the third power) by the inverse of an exponential function (the number e raised to the power of the variable hv / kT). Exponential functions get larger faster than power functions do, which means that Planck's function has a maximum intensity at any temperature T. Consider, for instance, the function $f(x) = x^3 / 2^x$, which is very similar to Planck's intensity functions. If we start plotting the values of $f(x)$ for $x = 1, 2, 3, \ldots$, we get $\frac{1}{2}$, 2, $3\frac{3}{8}$, 4, $3\frac{29}{32}$, and then the values of f start decreasing quickly toward zero.

Second, if hv is much larger than kT, $e^{hv/kT}$ is so large that subtracting 1 from it (as the denominator in Planck's formula tells us to do) hardly changes its value at all, so for these frequencies, $I(v,T) = (2hv^3 / c^2) / e^{hv/kT}$. Planck immediately recognized that this had the same form as Wien's approximation $I(v,T) = Av^3 e^{-Bv/T}$ (because $1 / x$ and x^{-1} mean the same thing). Moreover, the constants A and B in Wien's approximation, which Wien had obtained empirically, were now revealed to be constants with physical significance. The number A was $2hv^3 / c^2$, and the number B was h / k.

In order to appreciate the third attractive feature of Planck's formula, we need to use a result from calculus that has its antecedents in Zeno's Paradox. Zeno's Paradox is a conundrum that asks how an arrow can reach its target if it first travels half the distance to the target, then half of the remaining distance, then half of the remaining distance . . . and so on. It would seem that the arrow can never reach its target, as it always travels half of the remaining distance. However, if we think of the distance to the target as 1, the distance that the arrow travels in each of the stages of Zeno's Paradox is ½ in the first stage, ¼ in the second, ⅛ in the third, etc. The sum of the distances of all these stages is ½ + ¼ + ⅛ +. . . .

A resolution of Zeno's Paradox can be extended by looking at the more general problem of finding the sum of the geometric series $r + r^2 + r^3 + \ldots$, where r is a number between 0 and 1.

If the sum of these numbers is denoted by S, then $S = r + r^2 + r^3 + \ldots$. Multiplying both sides of this formula by r yields $rS = r^2 + r^3 + r^4 + \ldots$. When we subtract the infinite sum for rS from the infinite sum for S, since all the individual terms that appear in the infinite sum for rS also appear in the infinite sum for S, the only uncanceled term in S is the first term r. So $S - rS = r$. The left side can be factored, which yields $(1 - r)S = r$. Consequently, $S = r / (1 - r)$. In the case of Zeno's Paradox, $r = \frac{1}{2}$, and we breathe a sigh of relief as we see that $S = \frac{1}{2} / (1 - \frac{1}{2}) = 1$; the arrow really does reach its target.

In the eighteenth century, the techniques of calculus, notably those derived by the English mathematician Brook Taylor,[8] were used to obtain infinite sum descriptions (aka infinite series representations) of many functions. The sum above was seen as an infinite series representation of $f(r) = r / (1 - r)$. One of the most basic functions for which such a representation is available is the exponential function $f(r) = er$, which has the representation

$$er = 1 + r / 1 + r^2 / (1 \times 2) + r^3 / (1 \times 2 \times 3) + r^4 / (1 \times 2 \times 3 \times 4) + \ldots$$

In particular, for very small values of r, the first two terms of this series, $1 + r$, constitute an extremely accurate approximation to er. Planck, of course, was well aware of this, and in cases where hv is much smaller than kT, the denominator of his intensity function, $[1 / (e^{hv/kT} - 1)]$, could be closely approximated by $(1 + hv / kT) - 1 = hv / kT$. Substituting this into his expression for $I(v,T)$, Planck obtained $I(v,T) \approx (2hv^3 / c^2) \times kT / vh = (2k / c^2)v^2T$—the Rayleigh-Jeans Law!

Talk about pulling rabbits out of a hat! With the assumption that the oscillators only radiated energy in integer multiples of hv, Planck had derived a formula that circumvented the ultraviolet catastrophe

and reduced to both the Rayleigh-Jeans Law and Wien's approximation in the regions where both were known to be correct—to say nothing of unearthing the meaning of the mysterious constants in Wien's
approximation.

Even more rabbits were soon forthcoming—after all, what do you
expect from rabbits? (My parents could have answered this question:
when they sailed for Bermuda in 1935 on their honeymoon, one of their
more mischievous friends had arranged for a pair of male and female
rabbits to be placed in their stateroom as a wedding gift; they off-loaded
about ten rabbits when they reached Bermuda.) Recall that Kirchhoff
had shown that it didn't matter how you achieved thermal equilibrium;
substances and shape were irrelevant. Planck had used electromagnetic
oscillators to produce the radiation, and so when Boltzmann's constant
popped out of his derivation of the radiation law, it demonstrated a possible connection between electromagnetism and the not-yet-com-
pletely-accepted atomic theory.

Planck's comment to his son, that he had come up with an idea that
was potentially as important as those of Newton or Einstein, was indeed
prescient. The Nobel Prize Committee put Planck on the short list for
the prize in 1907 and 1908. In fact, he almost received it in 1908—not
for the quantum hypothesis that lay at the heart of his derivation, but
rather that his calculations helped confirm the atomic theory.[9] However,
he was not to receive that prize until 1918—but by then the Prize was
awarded for "the services he rendered to the advancement of Physics
by his discovery of energy quanta."[10]

Indeed, it took some time for the physics community to appreciate
that the concept of the energy quantum was truly the *pièce de résistance*
of Planck's theory. For a number of years, the energy quantum was
seen merely as a mathematical trick that simultaneously avoided the
ultraviolet catastrophe and reduced to Wien's approximation and the
Rayleigh-Jeans Law under the appropriate conditions. Mathematics is
the language of physics, but sometimes the connection between mathematical symbols and the real world is not apparent. Ideally, one would
like to have the mathematical theories that produce practical formulas
that accord with the real world to have been derived from assumptions

and observations about the way the real world is, rather than from a hypothetical surmise with no apparent connection to the real world.

What Is This Thing Called *h*?

The units in which a constant is expressed can often be determined from an equation in which that constant appears. If we look at Newton's formula for the gravitational force, $F = GmM / r^2$, for example, we can see that the value of G must be expressed as the product of mass units times the cube of distance units divided by the square of time units. This is because force—to use the metric system—is measured in kilograms × (meters / seconds^2) and the expression mM / r^2 is measured in kilograms^2 / meters^2. In order to have the same units appearing with the same exponents on both sides of the equation, the units for G must be kilograms × meters^3 / seconds^2.

The same logic applies to h—the equation $E = h\nu$ requires its units to be of the form energy-units × time-units, and the calculated value of h, using the units popular at the time, was 6.62×10^{-27} erg-seconds. An erg is a unit of energy; one of my physics professors used to describe it as approximately the amount of energy an ant needs to stamp one of its feet. I don't know how accurate this is, but it gives you the idea that an erg is a very small amount of energy, so 6.625×10^{-27} ergs is an inconceivably small amount of energy. An erg-second is the result of expending one erg of energy for one second. Ants probably don't stamp their feet, and if they do it's for a duration considerably shorter than one second, so—with apologies to my physics professor—since most of us have seen an ant labor continuously to push a grain of sugar or some similar substance, let's suppose that an erg-second is the energy expenditure of an ant pushing a grain of sugar for one second. As we have seen, visible light is radiated at a frequency on the order of 5×10^{14} cycles per second, so the lowest energy level of radiation for visible light is on the order of $h\nu = 6.55 \times 10^{-27} \times 5 \times 10^{14} = 3.28 \times 10^{-12}$ ergs. Consequently, you need about 300 billion such transitions to generate the same amount of energy as an ant uses in stamping one of its feet. The time required to emit a single photon varies, but it's on the order of a tenth of a nanosecond, 10^{-13} of a second. That means that you'd need about 3×10^{24} such transitions to equal the amount of effort expended by an ant pushing a grain of sugar for one second.

Energy quanta are extremely small in the real world. Red light oscillates at a frequency of about 4×10^{14} cycles per second, so $h\nu = 6.62 \times 10^{-27} \times 4 \times 10^{14} \approx 2.65 \times 10^{-12}$. Therefore the minimum change in energy in red light is about one four-hundred-billionth of an erg. Had it been considerably larger, it would have shown up in the real world. For instance, most of us have seen dimmer switches on lamps; as you turn a knob the light level gradually diminishes from maximum brightness until the light is finally out. If Planck's constant were a lot larger, we wouldn't be able to have such dimmer switches—lights wouldn't appear to dim gradually but instead would dim by noticeable jumps—they might simply have a few different brightness levels, such as the ones we see in the 50–100–150 watt bulbs, whose brightness is controlled by a switch but can only assume one of those three levels. A much larger Planck's constant would result in lights such as the 50–100–150 watt bulb described above being an inherent property of lights; gradual dimmer switches would be impossible. Of course, Planck's constant shows that, to the sufficiently sensitive "eye" (in this case, an electronic rather than a biological one), in the real world there are no dimmer switches; light intensity dims by jumps, but those jumps are so small—about one four-hundred-billionth of an erg—that our eyes perceive them as gradually dimming. In fact, our inability to perceive subtle discontinuity is responsible for many of the most important devices of our technological world; TV and computer screens present slightly different pictures hundreds or even thousands of times a second, producing the illusion of continuous motion.

The Triumph of the Quantum Theory

As we have observed, it took a significant amount of time before the importance of the quantum hypothesis was fully appreciated. In fact, it is rather surprising that Newton's theory of gravitation—which spread in the seventeenth century, with no form of communication other than person-to-person, mail, or books and journals—was much more rapidly assimilated into the body of science than Planck's theory, when we already had the means for communication at close to the speed of light.

It wasn't until Einstein's "miracle year" of 1905 that physicists began to realize that Planck's quantum hypothesis was much more than a mathematical trick that happened to work. During that year, Einstein published three remarkable papers, the first of which dealt with the problem of the photoelectric effect. The photoelectric effect had first been observed by Heinrich Hertz, the discoverer of radio waves, but was more accurately analyzed by Philipp Lenard in 1902. Lenard discovered that shining light onto certain metals resulted in the production of an electric current. Lenard compared the energy of the electrons in the current from two different light sources, a zinc arc and a carbon arc. Although the light from both sources was a mixture of light at various frequencies, the dominant frequency from the carbon arc was higher than the dominant frequency from the zinc arc—and the average kinetic energy produced by the light from the carbon arc was higher than the average kinetic energy of the light from the zinc arc.

One might also think that if one were to increase the power of the light, the energy of the electrons produced would increase, too. Lenard, however, showed that this intuitive conclusion was not the case, and the average energy of the electrons depended upon the frequency of the light used to produce them. For this startling result, Lenard was awarded the Nobel Prize in 1905, the same year that Einstein used Planck's quantum hypothesis to explain it. A certain amount of energy was needed to free an electron from the metal atoms. If the frequency of the incident light was such that hv (the energy of the light) was less than the energy needed to liberate an electron, there would be no current. The idea here is somewhat akin to a needed threshold such as melting temperature. A swimming pool full of water boiling at 373 kelvin has enough heat to melt iron. However, if you immerse a piece of iron in the pool, nothing happens. That's because the heat in the water isn't available at a sufficiently high temperature. Einstein won the Nobel Prize for realizing that the photoelectric effect, with the aid of Planck's quantum hypothesis, could be explained by a similar threshold effect. In case anyone hadn't paid attention to the mathematics in the paper, Einstein spelled it out. "Monochromatic radiation," he wrote, " . . . behaves in thermodynamic respect as if it consists of mutually independent energy quanta of magnitude hv."[11]

Other confirmations followed. In 1913 the Danish physicist Niels Bohr proposed a new model of the atom. The energy of an atom normally had a certain value, which Bohr described as the atom's ground state. The atom could absorb photons of certain energies; doing so would boost the energy of the atom—but again, only by specified amounts. This would boost the atom into an excited state. In one of these excited states, the atom could emit photons only having particular energies—which corresponded to specific frequencies via Planck's $E = hv$ formula. Specific frequencies meant specific colors; and Bohr's model of the atom explained the colors in the spectra of the various atoms. Just as Wien had developed an empirical formula for radiation intensity that Planck's formula explained, Bohr's model of the atom explained an empirical formula for the lines in the hydrogen atom's spectrum that the physicist Johann Balmer had developed.

Einstein was later to write, "This discovery was to become the basis of all twentieth-century physics and has almost entirely conditioned its development ever since."[12] He wrote those words in a chapter titled "In Memoriam Max Planck," decades after the 1920s and 1930s, which had seen Planck's quantum hypothesis totally transform our view of the universe, including the dual particle-wave nature of light and electrons; the impossibility of determining the position and momentum of a particle to an arbitrarily high degree; and the entanglement of the quantum states of particles, by which measuring one affected the state of the other.

These properties were so bizarre that almost a century later, explanations of what they say about the universe have yet to be offered that are completely satisfactory to the entire physics community. There have been almost as many books written about these phenomena as there are season-by-season analyses of the TV series *Lost*. I urge you to read one or two—or more—of them (the books, not the season-by-season analyses). As the British astronomer Sir Arthur Eddington wrote, "Not only is the universe stranger than we imagine, it is stranger than we can imagine."[13] It is even stranger than *Lost*.

CHAPTER 9

THE
SCHWARZSCHILD
RADIUS

I f I had but world enough and time—or if there is such a thing as
reincarnation—I would like to be an astronomer. Of all the sciences,
it has the most appeal for me—possibly because I can understand with
relative ease even technical articles on the subject. Of all the natural
sciences, astronomy has always impressed me as the one that is closest
to mathematics in the way the trade is practiced. Until recently, one
could not really perform experiments in either mathematics or astronomy the way that one can in physics, chemistry, or biology. The computer has changed that, of course, making simulations an integral part
of the advance of astronomy, and the exploration of space has brought
the planets and the Sun "up close and personal" in a way that could
never have been anticipated a century ago. Nevertheless, many of the
great advances in astronomy were made long before space exploration
and the modern computer, and they stand as a testament to the incredible reach of the human intellect. One of the finest examples is the
Schwarzschild radius, which describes the black hole left after the
death of a star. To fully understand it, however, we first need to talk
about a star's life.

With the exception of the Sun, the stars are almost inconceivably far
away. And ours—at 93 million miles away—isn't particularly close.

Even if the Sun, which has a radius of about 440,000 miles, were actually the size of a grapefruit with a radius of about three inches, the Sun would still be 53 feet from Earth (which, to match the scale, would have been shrunk to the size of a small BB). At actual scale, it takes light 8 minutes to travel from the Sun to the Earth, but 4.3 years to travel from Alpha Centauri, the next-closest star. Using our model, Alpha Centauri would be about 2,800 miles from the Sun; if the grapefruit were located at home plate in Dodger Stadium in Los Angeles, the Earth would almost be on the pitcher's mound and Alpha Centauri would be somewhere in New Brunswick, Canada.

As a result, astronomers prior to the year 1950 had only two tools with which to construct theories of stellar behavior. The first was an analysis of the electromagnetic radiation emanating from the stars, and the second was using the theories of gravitation, thermodynamics, electromagnetism, and nucleosynthesis to make predictions about how a big ball of gas ought to behave.

There is something truly awe-inspiring in our ability to understand so much about things that are so far away.

Carl Sagan and Stellar Taxonomy

There is a poignant passage in Carl Sagan's *Cosmos*, in which he describes how, as a small boy, he visited a local library and asked for a book about the stars.[1] The librarian found a book on movie stars of the era and gave it to Sagan, that being, of course, not what Sagan was looking for.

I read this passage when the book was published, in 1980, and it seemed a little sad, that the word *star* would be more quickly associated with celebrity than astronomy. But maybe it's something we can work with. I didn't get married until 2000, and when I did, it was to a woman who had a much greater interest in celebrities than I did; when we would board an airline I would take a recent copy of *Science* or *Scientific American*, whereas Linda steeled herself for the flight with a copy of *People*. I idly picked up a copy and thumbed through it one day, when it suddenly struck me—as it undoubtedly has struck others—that there are a lot more parallels between the two meanings of *star* than I had originally suspected (even though I fervently hope that *star* will always first refer to an astronomical object).

First of all, exactly what is a celebrity? The definition I find most reasonable is someone who is widely recognized. Of course, the term "widely" is somewhat vague, but I would think that a person who is known considerably beyond the circle of individuals he or she has personally encountered would qualify. If you are somewhat cynical, you might observe that the more celebrity people have, the more likely they are to feel that their personal opinions are of great importance: thus the first similarity between a star and a celebrity is that the former is a huge mass of overheated gas, and the latter has a tendency to become one.

Stars and celebrities share two more important attributes. Luminosity can describe one of them. For a star, luminosity is the rate at which it radiates energy, either visibly (to us) or entirely. Apparent luminosity measures the visible energy, and bolometric luminosity (named for the instrument devised to measure this quantity) is the total. Although it is not so precisely defined in dealing with celebrities, a celebrity's luminosity can be described as how recognizable that celebrity is.

We can also describe a star's color, which we know to be an indication of its temperature. The analogy for the celebrity would not be the physical color imparted by skin pigment but a measure of temperature accorded by how much attention is paid to a celebrity by the news media. There is a rationale for using temperature to describe this, as it is increasing media coverage that puts the heat on celebrities, as Tiger Woods or Mel Gibson could attest. We now have two different measuring scales for both types of stars: temperature and luminosity. We can now draw two diagrams—one for stars and one for celebrities—using these two scales as horizontal and vertical axes, and locate each star or celebrity on the corresponding diagram by plotting a point with the appropriate temperature and luminosity coordinates.

The Temperature-Luminosity Diagram

We can make a graph of temperature and luminosity, in which news media recognition forms the horizontal scale and public recognition the vertical scale. On this graph, public recognition will increase as we move upward, but news media recognition will *decrease* as we move from left to right. Points—each indicating a specific celebrity—on this diagram are not randomly scattered with no apparent pattern. For starters, there is a significant positive correlation between public and news media recognition of the celebrity: generally, the more the news media recognizes a celebrity, the more the public will as well, and vice versa. A large number of celebrities will therefore be positioned on a band that stretches from the upper left of the diagram—high public and news media recognition—to the lower right—low public and news media recognition (sadly, where most of the people described in this book are located, excepting outliers such as Albert Einstein). But the correlation is not strict: there are large groups of celebrities with high news media recognition and little public recognition. Many polls have revealed that although the vice president probably gets covered by the news media every day, a depressingly large segment of the population does not know his name and of those who recognize the name Joe Biden, very few would recognize him on the street (I'm one of them). There are also groups of celebrities who are readily recognized by large segments of the public but receive little news media coverage; many entertainers fall into this category.

This diagram can also be used for plotting the career arc of a celebrity. For a particular celebrity, we could place one point on the diagram corresponding to the celebrity's position for each year of the celebrity's life. We could then connect the dots in chronological order; the resulting path would describe how the celebrity's recognition has changed with time. Some celebrities, such as Marilyn Monroe, initially appear on the bottom right of the diagram, only to catapult to fame and high recognition by both the public and the news media, and remain there, even long after death.

We can do the same for actual stars. The original temperature-luminosity diagram was constructed in the second decade of the twentieth century by two astronomers, Ejnar Hertzsprung and Henry Norris Russell. This diagram had temperature decreasing on the horizontal scale—corresponding to the same shifts in color that had been studied by Boltzmann at the end of the nineteenth century—and luminosity increasing on the vertical scale.

I must admit when I first saw the Hertzsprung-Russell diagram,[2] I wondered why temperature decreased as one went from left to right. Both Denmark and the United States, the countries of which Hertzsprung and Russell were citizens, have languages that are read left to right, and it is traditional for scales to increase from left to right on graphs. However, the relation $kT = h\nu$ tells us that temperature T increases with increasing frequency ν of light—but the frequency of light is inversely proportional to the wavelength λ of the light, which is defined as the distance between two successive crests of the wave. If the frequency of light doubles, so that twice as many complete cycles appear in a fixed period of time, the distance between successive crests must halve. Both frequency and wavelength are natural ways to describe light, and my guess is that Hertzsprung and Russell originally used wavelength as a horizontal axis. Then, as the diagram emerged, they found a reason to switch from wavelength to frequency as the horizontal axis.

Each star corresponded to a point on the diagram, and once graphed, patterns emerged that were to initiate studies into the life cycle of stars.

The first pattern was the emergence of a large band from the top left of the diagram to the bottom right—similar to the pattern described earlier in the celebrity temperature-luminosity diagram. This band is called the main sequence, and the majority of stars reside on it. However, there are two correlations involving the main sequence that do not exist for the corresponding band in the temperature-luminosity diagram for celebrities. As one moves down the main sequence, from the upper left to the lower right, the stars not only become redder and less luminous, but also less massive and longer-lived. A star on the upper left of the main sequence can be sixty times as massive as the Sun, but will also live only 10 million years or so. In contrast, a star on the lower right of the main sequence may weigh less than one-tenth the mass of the Sun, but could survive for more than 100 billion years.

There are several groups of stars that do not appear on the main sequence. We have encountered one of these groups before: the white dwarves cluster in the middle of a band that would be parallel to the main sequence but some distance below it. Roughly the same distance above the main sequence as the white dwarf band is below it, and also roughly parallel to the main sequence, are the giants and supergiants, the group of stars that we will focus on in this chapter.

One can also trace the life of a particular star as a curve on the Hertzsprung-Russell diagram, similar to the career arcs on the temperature-luminosity diagram for celebrities. There is, however, an important difference between these two. Of course, we cannot actually follow a star through its life cycle to plot such a curve; even the shortest-lived stars last far longer than human history. However, the life cycles of stars are completely determined by their temperature and luminosity at any moment in time—if we find two stars with the same temperature and luminosity, their lives will follow the same paths. This characteristic is very different for celebrity arcs—celebrities who have the same temperature and luminosity coordinates may follow very different career paths: one may go on to greater glory, one may crash and burn, and a third may simply sputter out.

The fact that the life of a star is completely determined by two coordinates has a familiar parallel in an earthly phenomenon—the path of a projectile. If we place an old-fashioned cannon (little more than an iron tube that is closed at one end) on the ground and fire cannonballs with the same weight but use differing amounts of gunpowder, the amount of gunpowder completely determines the path of the cannonball, and no two paths corresponding to different amounts of gunpowder will intersect. If we are able to accurately determine the position of a cannonball at just a single moment, whether in the air or when it hits the ground, we know exactly how much powder was used, where the cannonball was, and where it will be at all moments in time (of course, we're always referring to cannonballs fired from the same cannon). The similarity between the life cycles of stars and the paths of cannonballs is a result of the fact that both sets of curves—the life-cycle curve of a star on the Hertzsprung-Russell diagram and the curve of a cannonball in space—are solutions to differential equations, a class of equations that arises through calculus by looking at the rates at which various parameters are changing. It is a characteristic of many differential equations, including the ones governing projectile motions (originally discovered by Newton) and the energy balance in stars, that the solutions form a collection of nonintersecting curves.

It is the mass of the star that determines its luminosity and temperature. A ball of hydrogen must be massive enough (about seventy times the mass of the planet Jupiter) that gravitational contraction can elevate temperatures to the point where hydrogen fusion commences. The dynamic mechanism of a star is fairly simple. The star contracts from the force of gravity, which creates high pressures at the center of the star, which result in high temperatures—just as the ideal gas formula predicts. Energy flows from the hot interior to the cooler surface, where it is radiated away. Actually, this process does not depend on thermonuclear fusion; the same thing would happen if the Sun were made of coal, as Kelvin hypothesized when he showed that chemical combustion could not sustain the Sun for long. (We are actually fortunate that the Sun is not made of coal, because if it were, the pressure at the core would be vastly greater, producing much greater temperatures at the interior—and we would fry in an instant.)

The upper limit on the possible size of a star was initially determined by Arthur Eddington, who calculated that if a star is about 120 solar masses, the outward radiation pressure would overwhelm the inward force of gravity.[3] Either the star would blow itself apart or expel enough mass to get down below that limit. (Although this calculation is certainly correct in principle, in 2010, a paper was published in which the star R136a1 was determined to have a mass more than 250 times that of the Sun.[4] If this mass is confirmed, Eddington's calculations will have to be revised.)

We can be sure of one thing, however. The supergiants—the truly massive stars—are destined for extraordinary fates, and they are the ones that, as Sagan said, make us star-stuff. These are the most intriguing stars—they burn unbelievably hot and have extremely short lives, at least for stars. This, too, parallels the nature of celebrity—big celebrities of my era have been mostly forgotten, but the ones that burned white-hot and died young—Elvis and Marilyn Monroe—are still widely recognized and have their own cult following.

The Great Technetium Mysteries

One of the most interesting bits of star-stuff is something that, unless
you are a chemistry geek, you've probably never heard of, even though
you may have come into contact with it, as it is in fairly widespread
use as a radioactive tracer. This bit of star-stuff is technetium, element
number 43. Technetium fills one of the three gaps in the initial version
of the periodic table as propounded by Mendeleyev. One of the hallmarks of a great theory is its ability to make surprising predictions.
The periodic table certainly meets this requirement, as Mendeleyev
correctly predicted some of the physical and chemical properties of element number 43, which he initially called eka-manganese, because it
lay directly below manganese in the periodic table. What he could not
have predicted was the fact that it was radioactive, as radioactivity was
unknown at that time.

In fact, there are no nonradioactive isotopes of technetium, and it is
incredibly rare on Earth. It is the first element to have been produced
by artificial means: hence the name "technetium" from the Greek word
for "artificial." Its existence was discovered, by Emilio Segré, in
molybdenum foil from a discarded radioactive cyclotron. It was soon
discovered that the great majority of technetium's isotopes are extremely short-lived, but there are a couple with half-lives (the time it
takes for half a quantity to radioactively decay) of several million years.

How we could know such a thing puzzled me as a child. It's not as
though, I reasoned, anyone could possibly have been around long
enough to measure it! The answer lies in calculus.

Imagine you have two buckets of technetium, one containing twice
as much as the other. If you leave them alone for some time and then
measure the amount of technetium in each, you will find that the first
bucket still contains twice much as the second, even though the amount
in both containers has decreased. The rate of change is proportional to
the amount present.

The functions that solve the differential equation in which the rate
of change of a substance is proportional to the amount of substance
have the form $f(t) = N \times 2^{t/h}$, where t is the amount of time that has

elapsed since the initial measurement was made. N is simply the amount of substance that is present at the time of the initial measurement (corresponding to $t = 0$); h is positive for things that increase with time, such as colonies of bacteria, and negative for things that decay with time, such as radioactive substances. The absolute value of h is the half-life of the substance.

Suppose we initially use a Geiger counter on a radioactive substance and discover that the click rate is 1,000 clicks per minute. We measure the same substance precisely 100 hours later, and discover that the click rate is 999 clicks per minute. Because click rate is proportional to the amount of radioactive substance present, we can simply assume that our unit of mass is such that a single unit of mass would emit one click per minute. Therefore our initial quantity of mass $N = 1{,}000$ and $f(100) = 999$. But then $999 = f(100) = N \times 2^{100/h} = 1{,}000 \times 2^{100/h}$. So $0.999 = 2^{100/h}$, and we can solve for h by using logarithms. It doesn't matter whether we use common (base-10) logarithms or natural (base-e) logarithms, but since common logarithms are more widely known, I'll use them. So, by a well-known property of logarithms

$$\log 0.999 = \log (2^{100/h}) = (100 \, / \, h) \log 2$$

Multiplying both sides by h and subsequently dividing both sides by $\log 0.999$ shows that

$$h = 100 \log 2 \, / \log 0.999 = -69{,}280.1 \text{ (in hours)}$$

Since h is negative, its absolute value is the half-life of our substance, approximately 7.91 years. We could even determine what the substance is by looking at a table of half-lives; there are probably only one or two substances with a half-life of precisely 7.91 years.

Indeed, technetium's very short half-lives prove that most of the elements we see were made in stars. The longest-lived isotope of technetium has a half-life of about 4 million years, and it is formed from the radioactive decay of much heavier elements such as uranium. Every 4 million years, half of it disappears, so after a billion years a quantity of technetium has been cut in half 250 times. Since $(\frac{1}{2})^{250}$ is on the order of 10^{-76}, and since there are roughly 10^{80} atoms in the universe, of which very few are technetium, the fact that technetium can be found in the atmospheres of stars that are considerably older than a billion years is evidence that technetium is being minted in the stellar core. This constitutes *prima facie* evidence that our understanding of the processes that go on in the stellar core is correct. Since the contention that we are all star-stuff depends critically on our understanding of these processes, it's nice to get confirmation of our theories.

The Worst Experimentalist in History

In high school, I thought that was a good description of Nancy, my lab partner in chemistry. A big plus about Nancy was the fact that she had access to cigarettes in an era in which it was deemed cool to smoke, so we reached an agreement: I'd write up the experiments, and she'd slip me cigarettes (wild behavior in high school was considerably milder in the 1950s than it is now). The downside was a real risk of injury. I still remember one day when Nancy added 30 milliliters of concentrated sulfuric acid to sodium bromate rather than the 3 milliliters of dilute sulfuric acid. The result was that an ominous orange cloud of bromine started to materialize, from which, fortunately, our teacher saved us.

It turns out, however, that Nancy was a considerably better experimenter than Wolfgang Pauli, whose mere presence in laboratories was thought to be able to adversely affect any experiment in progress. But Pauli was a brilliant theoretical physicist, and enunciated a concept— the Pauli Exclusion Principle—which explained the mechanism that enabled the formation of the white dwarves.

The Pauli Exclusion Principle is a fundamental concept in quantum mechanics. It states that no two members of a class of particles called fermions, which includes electrons and quarks (and the common composites of them, such as neutrons and protons), can have the same quantum state. A quantum state is an aggregation of quantum properties, one of which is energy level. A consequence of this is that electrons that are close together must have different energy levels. Electrons typically occupy low-energy levels, but if there are a lot of them packed really close together, some of the electrons must occupy high-energy states. These electrons create a type of pressure known as electron degeneracy pressure (which we will encounter again in Chapter 11). Unlike the pressure in the Ideal Gas Law, this pressure is a quantum-mechanical effect, and is not sensitive to temperature. This leads to interesting effects inside a star—the degeneracy pressure, in concert with radiative pressure from fusion, prevents gravitational collapse. Fusion, however, eventually stops. In some stars, which are known as white dwarves, what remains is usually carbon and oxygen, which glow because they are hot, and which are supported against gravity by electron degeneracy pressure. Larger stars, however, can undergo further fusion, but even they must stop at iron, as the fusion of iron absorbs energy rather than producing it, as does fusion of elements lighter than iron. In the absence of fusion's radiative pressure, gravity—if the star is large enough—overwhelms electron degeneracy pressure. In approximately one-tenth of a second the gravitational collapse occurs at about 25 percent of the speed of light. The electrons are squashed into the protons, with the result that the entire star consists of neutrons. Several solar masses are now squished into a sphere that is perhaps ten miles in diameter. The resulting shock wave rebound tears the outer layers of the star apart, and a supernova appears in the sky. In this instant, the star releases a hundred times as much energy as the Sun will generate in its entire lifetime. Most of this energy is released as neutrinos.

Cosmic Gall

Matter is virtually transparent to neutrinos—as the poet John Updike immortalized in the poem "Cosmic Gall":

> *Neutrinos, they are very small.*
> *They have no charge and have no mass*
> *And do not interact at all.*
> *The earth is just a silly ball*

To them, through which they simply pass,
Like dustmaids through a drafty hall
Or photons through a sheet of glass.
They snub the most exquisite gas,
Ignore the most substantial wall,
Cold-shoulder steel and sounding brass,
Insult the stallion in his stall,
And scorning barriers of class,
Infiltrate you and me! Like tall
And painless guillotines, they fall
Down through our heads into the grass.
At night, they enter at Nepal
And pierce the lover and his lass
From underneath the bed—you call
It wonderful; I call it crass.

Updike would call it a lot more than crass if the Sun were to explode in a supernova (don't worry, it's not massive enough). Even though a single neutrino can pass through a lead barrier a light-year thick without interacting with one of its atoms, the neutrinos generated by a supernova of our Sun would be so energetic and so numerous that the radiation could kill a human being as far away from the Sun as Jupiter is.

Incidentally, there is a small technical error in Updike's poem—neutrinos actually have mass, although not a whole lot. Updike passed away in 2009, and it's a pity he didn't choose to write more poems about neutrinos, as there are a number of enchanting mysteries surrounding them that I think would delight a poet. There are three different types of neutrinos, and apparently they can change their type in mid-flight.[5] Gender change among the neutrinos! Someone should have brought this to Updike's attention.

What happens next (to the supernova, not Updike) depends upon the mass of the original star. If the remnant neutron core is less than about 2.5 solar masses, it continues to exist as a neutron star. A cubic centimeter of the material in the center of a neutron star weighs an astounding 10^{15} pounds. The gravitational force at the surface of the star is more than 100 billion times the strength of the gravitational force on Earth, but if what remains of the star weighs less than about 2.5 solar masses, neutron degeneracy pressure will counterbalance even this crushing force, and the neutron star will continue to exist. If what remains of the star weighs more than 2.5 solar masses, though, even neutron degeneracy pressure cannot counterbalance the gravitational force, and the star disappears from the universe in the form of a black hole.

A black hole is aptly named, for it is a region of space in which the gravitational force is so intense that not even light can escape. Although black holes first entered public consciousness around 1967, when the physicist John Archibald Wheeler coined the term, the concept of a black hole goes back more than two centuries—to John Michell, the man who lent Henry Cavendish the torsion balance with which Cavendish determined the density and weight of the Earth. In fact, Michell wrote to Cavendish, "If the semi-diameter of a sphere of the same density as the Sun were to exceed that of the Sun in the proportion of 500 to 1, a body falling from an infinite height towards it would have acquired at its surface greater velocity than that of light, and consequently supposing light to be attracted by the same force in proportion to its *vis inertiae*, with other bodies, all light emitted from such a body would be made to return towards it by its own proper gravity."[6] This is pretty sensational stuff, as Michell not only predicted photons (that stuff about light being attracted by the same force), but even worked out to some extent the mathematics of a black hole.

Michell's work, however, did not attract much attention, and it took a German soldier stationed on the Russian front during World War I to regenerate interest in the black hole phenomenon.

Karl Schwarzschild

I was fortunate enough to have been too young for either of the two World Wars or the Korean conflict, and even though I was of draft age during the early days of Vietnam, the U.S. government decided that it was more in the national interest for me to study mathematics than to serve in the military. I duly reported for the physical (which I passed)

and the Army Alpha intelligence test. When the examiner found out
that I was a graduate student at Berkeley, he said that he expected me
to set records—at least for that particular induction center. He was
probably disappointed. The first two parts of the test were vocabulary
and arithmetic; as might be expected I nailed them. The third part was
spatial relations; you were shown an oddly shaped diagram with dotted
lines and were expected to identify in a multiple-choice format what
the diagram would look like if it were folded along the dotted lines.
This isn't one of my strong suits, as my spatial perception is not very
good. The third part was a breeze, however, compared with the fourth,
in which you were shown a tool and asked to identify what it was used
for. Since none of the tools was a hammer or a saw, I'm not sure I got
any of them right. At any rate, I spent most of the Vietnam War teaching
and studying math, and am extremely grateful I did not have to see active service, as I am sure I would have been able to think of nothing
but how long it would be until I could go home.

Karl Schwarzschild was made of much sterner stuff than I am. Einstein published his General Theory of Relativity in 1915, and while
Schwarzschild was on the Russian front, he not only managed to obtain
a copy to study, but did significant research as well. The theory is expressed as a system of differential equations (which I think of as the
language of the universe, because they occur so frequently throughout
the sciences), and Schwarzschild was the first to obtain specific solutions to those equations, which he communicated to Einstein. Einstein
thought so much of the work that he presented them personally to the
Prussian Academy of Sciences, which then published them.

Sadly, Schwarzschild died in 1916 of an autoimmune disease contracted while serving on the Russian front. World War I was notable in
that it may have been the last war in which the combatants acknowledged the humanity of those on the other side. The death of Paul
Ehrlich, the father of chemotherapy, was acknowledged by the entire
world as the passing of an individual who had contributed inestimably
to the betterment of the human condition. Of Schwarzschild's passing,
Eddington said, "The war exacts its heavy toll of human life, and science is not spared. On our side we have not forgotten the loss of physi-

cist Moseley, at the threshold of a great career; now, from the enemy, comes news of the death of Schwarzschild in the prime of his powers. His end is a sad story of long suffering from a terrible illness contracted in the field, borne with great courage and patience. The world loses an astronomer of exceptional genius."[7]

Schwarzschild originally phrased his results in the framework of the General Theory of Relativity, but it is also possible to obtain the basic idea using simple Newtonian physics, as Michell must have done. To do so, imagine that a mass M is concentrated in a nonrotating sphere of radius R. If we fire a projectile of mass m up from the surface of the sphere with velocity v, it will fail to escape the gravitational pull of the sphere if the kinetic energy of its motion, $\frac{1}{2}mv^2$, is insufficient to counteract the gravitational potential energy GMm / R exerted by the sphere on the projectile. The fastest that a projectile could travel would be c, the speed of light. So, if $GMm / R > \frac{1}{2}mc^2$, even light cannot escape from the sphere. Notice that we can divide both sides of this inequality by m; if after doing this we solve for R, we see that $R < 2GM / c^2$. So if the mass M lies inside a sphere of radius $2GM / c^2$, no light (and no information) can escape from the sphere. The quantity $2GM / c^2$ is called the Schwarzschild radius, and the surface of the sphere centered at the center of the mass M and whose radius is the Schwarzschild radius is called the *event horizon*. As far as we can tell, no events take place inside the event horizon, but because no information gets out to us from inside the event horizon, there may be the remote possibility that there is one hell of a party going on.

Unlike all the other constants we have examined in this book, it is not a constant in the strict sense of the word, as its value depends upon the mass M. The Schwarzschild radius of the Earth (or, more precisely, of a mass equal to that of the Earth) is about one centimeter, and the Schwarzschild radius of the Sun is about 3 kilometers.

Black holes are conventionally portrayed as ominous black spheres that we imagine are incredibly dense, far denser even than a neutron star. Although a black hole that has shrunk to a point has infinite density, if this even makes sense, the density of massive black holes is surprisingly low. A galaxy may contain on the order of 10^{42} kilograms; if so its Schwarzschild radius is about 10^{15} meters. The volume of such a sphere is about 4×10^{45} cubic meters, so the density of the black hole would be about $\frac{1}{4000}$ kilograms per cubic meter, or about $\frac{1}{4}$ of a gram per cubic meter. At sea level air weighs about 1,200 grams per cubic meter, so the atmosphere is about 5,000 times as dense as a galaxy-sized black hole.

Black holes have been theoretical constructs for some time, but the evidence for their actual existence has been accumulating for forty years, and the star Cygnus X-1 seems to have all the characteristics needed for a black hole.[8]

One Hell of a Party

I mentioned earlier that the Schwarzschild radius is not a constant in the strict sense of the word, as different masses have different Schwarzschild radii. However, there is one Schwarzschild radius that strikes me as being an absolute constant, and that is the Schwarzschild radius of the universe.

There are currently reasonably good estimates for the total mass of the universe (scientists are confident that it is known to a factor of about five), and if we compute the Schwarzschild radius for it, we arrive at a number between ten and one hundred billion light-years. The universe is estimated to have a radius of about thirteen or fourteen billion light-years, so it is still inside its event horizon—and yes, there is one hell of a party going on. But the universe is still expanding, and at some time in the future, it is possible that the universe will expand beyond its Schwarzschild radius. Or will it? I've queried several physicists about this question, and haven't obtained a satisfactory answer. Maybe the universe hits its Schwarzschild radius and bounces inward, mirroring in reverse the shock wave rebounding off the stellar core in a supernova. Or something else. I won't be around to see it, but maybe I'll live long enough that I get to find out what will happen.

CHAPTER 10

THE EFFICIENCY OF HYDROGEN FUSION

My first year in graduate school, a country singer named Skeeter Davis took the song "The End of the World" to number two on the charts by asking one plaintive question: "Why does the Sun go on shining?"[1] The song shows it's tough to beat a broken heart for seeming bigger than the biggest questions. But it also shows that you can't trust country singers—or country music songwriters—to ask the great questions. The great question is not *why* the Sun goes on shining—it's *how*. I imagine people have been asking it since the beginning, but it wasn't until the nineteenth century that physics had derived the computational tools needed to attack the question. The answers that nineteenth-century physicists obtained would eventually lead to the number .007, which is in some respects the most important number in this book. As we shall see, had this number been a little lower or a little higher, this book—and "The End of the World," and all other songs and books—would never have been written.

Solar Power

Lord Kelvin was probably the first person to attack the problem of *how* the Sun goes on shining from a scientific standpoint. He started by calculating the power that the Sun generates. Measuring the sunlight incident at the top of the atmosphere, we find that every square meter receives 1.3 kilowatts. (It must be done there, rather than at the surface, because the atmosphere—thankfully—absorbs some of that power before it hits the Earth, otherwise they'd probably be wearing swimsuits in the Arctic.) The Sun is obviously radiating energy equally in all directions, and so we now calculate the amount of energy that hits the surface of a sphere whose radius is equal to the distance of the Earth from the Sun. That radial distance is 1.5×10^{11} meters, and since the surface area of a sphere with radius R is $4\pi R^2$, we see that a spherical shell whose radius is 1.5×10^{11} meters receives $4\pi \ (1.5 \times 10^{11})^2 \times 1.3 = 3.7 \times 10^{23}$ kilowatts of energy from the Sun. Power is the rate at which energy is produced; producing a kilowatt of energy for a second constitutes a kilojoule of power. If the Sun produced energy by burning a chemical fuel such as octane, it would generate on the order of 300 kilojoules per mole of fuel. The Sun weighs about 2×10^{30} kilograms, and octane, which has the chemical formula C_8H_{18}, has a molecular weight of 114. So a mole of octane weighs 114 grams, or 0.114 kilograms, and the Sun, if it were made of octane, would contain about 2×10^{30} / $0.114 = 1.75 \times 10^{31}$ moles. Burning the entire Sun would generate about $1.75 \times 10^{31} \times 300 = 5.3 \times 10^{33}$ kilojoules, which would run the Sun for about 5.3×10^{33} / $(3.7 \times 10^{23}) = 1.4 \times 10^{10}$ seconds. That's about 500 years. Even assuming the Sun were combusting hydrogen chemically, the Sun would burn for about 50,000 years, using these figures. Even before the twentieth century, there was evidence that the Earth had been in existence for hundreds of millions of years, so however the Sun was shining, it wasn't doing so by ordinary combustion.

After working this out (although Kelvin used coal as the fuel rather than octane or hydrogen), Kelvin looked around for another source of energy. He found it in the energy derived from the potential energy of the Sun's mass being converted to kinetic energy as it fell in to the center of the Sun. This would have heated the Sun for tens of millions of years—but that still wasn't good enough.

So how does the Sun go on shining? Much of the groundwork was laid during one of the most exciting periods in the history of physics, the decade between 1895 and 1905, with three revolutionary discoveries that would open the world within the atom. The final pieces of the puzzle would not fall into place until just before the start of World War II.

The Green Twig Fracture

Like most children, I didn't pay a whole lot of attention to my health. Other than an asthmatic episode at age three that required an injection of adrenaline, which I do not remember at all, I managed to escape most childhood medical problems. I still have my appendix and tonsils, and after I got over the mumps, my parents took me to Atlantic City, so I regarded the experience *in toto* as showing a profit. True, there was the annual trip to the doctor's office, which featured the 1940s vintage blood test—an infernal device that poked a hole in your finger by what seemed like a harpoon; even though only a few drops were smeared on glass slides for examination, it was far more uncomfortable than having a few syringes of blood drained from the vein in your arm. So I remained relatively unscathed until my parents got an invitation to bring the family to visit some friends who owned horses.

I clearly had done an inadequate job of communicating to my parents that I absolutely loathed the horseback riding lessons they had supplied as part of my childhood. Horses were a lot bigger than I was, and were very capricious. Nonetheless, I was a dutiful child, and so when my parents suggested that I go for a ride, I mounted one of the horses. A short while later, I must have inadvertently nudged it from second gear into third or fourth gear, for the horse suddenly picked up speed and I slid off the rear end of the horse. I put out my left arm to break my fall, experienced some pain at impact, but was well enough to get back on the horse and complete the ride.

My left arm swelled up somewhat overnight and was also a little sensitive, so my parents took me to a local hospital to have it examined. The doctor took an X-ray, showed it to us, and told us that I had a "green twig" fracture. You could see it perfectly on the X-ray; a bone in my arm looked like what happens when you try to break a green twig—it doesn't snap cleanly but breaks partially, with the two segments joined by portions of the twig that have frayed and partially separated but are still connected.

This took place in early June, and my arm was in a cast for a month—in the warm and humid Illinois summer. I experienced very little pain while it healed, but that month seemed to last forever. As Ogden Nash (maybe not the most celebrated poet of the twentieth century, but easily the most entertaining) so aptly put it, "One bliss for which there is no match, is when you itch to up and scratch."[2]

Maybe science was more impressive in the mid-twentieth century than it is today, because the difference between the things that science brought into one's life and ordinary day-to-day living were much more marked than they are today. The X-ray of my arm stood out in sharp contrast to most of the ordinary paraphernalia of everyday life, such as books and bicycles. A few years ago, my mother-in-law needed an MRI of her shoulder. It was far more sophisticated than the ghostly X-ray of my green twig fracture, but everyone took that sophistication in stride. Ho-hum, there's a lot flashier stuff available at the click of a mouse on the Internet.

Other children of my generation probably had a similar experience, became fascinated by how the green twig fracture actually healed, and went into medicine. I had absolutely no interest in how my arm was healing. I just wanted it to heal as soon as possible so I could get that damned cast off and experience the bliss for which there is no match. However, I was fascinated by the X-ray: how was it possible to obtain a picture, even though a shadowy one, of something that was invisible to the naked eye?

Science advances on at least two fronts. One consists of coming up with new explanations for well-known phenomena, such as when Newton explained the motion of the planets and Planck accounted for the intensity curves. The other front is the discovery of new phenomena. Philipp von Jolly, the man who told Planck all that remained in physics was to fill in a few missing details, was in retrospect doubly wrong. Not only did filling in missing details, such as Planck was to do, open up entire new vistas for exploration, but there were new phenomena to be discovered that would require explanations that nineteenth-century physics was completely unable to provide. One of these phenomena made it possible for the doctor to diagnose my green twig fracture some sixty years later.

Some scientific discoveries are the result of good fortune—Alexander Fleming's discovery of the effectiveness of penicillin mold in fighting bacterial infection is a good example[3]—but as Branch Rickey, a former general manager of the Dodgers said, luck is the residue of design.[4] By the end of the nineteenth century, scientists had not yet discerned the nature of electricity. They had a very good idea of how electricity behaved in metals and a reasonable idea of how electricity behaved in liquids—but the behavior of electricity in gases was poorly understood, and hence was the subject of considerable investigation.

Wilhelm Roentgen

If there had been a betting pool in the early 1890s on who would win the first Nobel Prize, Wilhelm Roentgen would have been the darkest of horses. He was expelled from the late nineteenth-century German equivalent of a high school for refusing to snitch on a fellow student who had drawn a picture denigrating a teacher. The German school system played hardball in those days—not only was Roentgen kicked out of the school he attended, but he was unable to continue his education in any German or Dutch school. I guess the German authorities weren't enamored of the philosophy of letting the punishment fit the crime. However, Roentgen executed an end run around this by enrolling in a Swiss university. There must have been a statute of limitations on failing to turn stool pigeon, as Roentgen eventually made his way back to German academia, where he spent the bulk of his career in the minor leagues of German universities—first Hohenheim, then Giessen, and finally Würzburg—doing absolutely nothing of note. Until November of 1895.

Scientific discoveries are often the result of improvements in technology—without the microscope, Anton von Leeuwenhoek would have remained a minor clothier in Holland rather than the discoverer of the world of the microbes. During the nineteenth century, better and better equipment had evolved, both to produce electricity, and to study gases under low pressure. One such development was the Crookes tube,[5] a glass container that would subject enclosed gases to high-voltage discharges of electricity. The electricity would be produced at the cathode, and sometimes fluorescence, the emission of light, could be observed at either the anode or from the walls of the glass tube. Producing fluorescence was a somewhat hit-or-miss process, depending upon a large number of variables, including the type of gas, the geometry of the Crookes tube, and the voltage of the electrical discharge.

Roentgen was fifty, well past what Newton described as the prime age for invention, when he had the good fortune to have a sheet of paper painted with barium platinocyanate, a substance known to fluoresce under ultraviolet light, lying near a Crookes tube. He applied voltage to the Crookes tube—and the barium platinocyanate fluoresced! Roentgen spent the next seven weeks in complete secrecy investigating this new phenomenon. One of the things that Roentgen observed was that the rays registered on photographic film. Two weeks into his investigations, he took the first X-ray photograph of his wife Anna Bertha's hand. On seeing the skeleton of her hand, she exclaimed, "I have seen my death!"[6]

Seven weeks later, as 1895 was coming to a close, Roentgen published a paper titled "On a New Kind of Rays."[7] Roentgen named the rays that were producing the fluorescence X-rays—X being the mathematical symbol for something unknown. On the second page of the paper was a description of the X-ray photograph of his wife's hand: "If the hand be held before the fluorescent screen, the shadow shows the bones darkly, with only faint outlines of the surrounding tissue."[8] As might be expected, this discovery, with its huge promise for revolutionizing medicine, brought forth new opportunities for Roentgen. He accepted a professorship at Munich in 1900—moving up to the major leagues—and in 1901 was awarded the first Nobel Prize for Physics. He donated the prize money to his university, and refused to take out patents on his discovery, even though they would have made him extremely wealthy, as he wanted mankind as a whole to benefit from his discovery.

Although Roentgen published three papers on X-rays between 1895 and 1897, by the time that he won the Nobel Prize, the nature of X-rays remained unknown. Roentgen was not to contribute further to the problem; his 1897 paper was his last, and he did not take an active part in scientific investigation while in Munich. It would fall to Max von Laue to discover the nature of X-rays, and to many others to speculate why Roentgen had stopped producing. Von Laue, who had a chance encounter with Roentgen in the third-class compartment of a train, had his own thoughts on the matter.

> Often one has asked for reasons why this man, after his epoch-making achievements of 1895–96, had so retracted. Many motives were suggested, some little flattering for Roentgen. I consider them all false. In my opinion, the impression of his discovery had so overpowered him that he, who made it when he was fifty, never recovered. For—and only few think of it—a great feat is a burden for him who achieved it. . . . It needed much to compile three papers, which, like Roentgen's from 1895 to 1897 exhausted the subject so much, that for a decade hardly anything new could be said about it.[9]

I have my own thoughts on the matter. I'm a baseball fan, and there's an obvious person in its history with whom to compare Roentgen— Don Larsen, a pitcher whose career record was an undistinguished 81–91 with a 3.78 ERA, yet who pitched the only perfect game ever in a World Series. Perhaps Roentgen was the same, a journeyman "player" who nevertheless achieved absolute greatness, for one brief period of his life.

What Laue discovered in 1912 was that X-rays were another form of electromagnetic radiation, of a substantially higher frequency than visible—or even ultraviolet—light. Visible radiation, as we have noted, has a frequency on the order of 5×10^{14} cycles per second. X-rays, however, have a frequency on the order of 10^{18} cycles per second. By Planck's $E = hv$ formula, we can see that X-rays have more than a thousand times the energy of visible light. This enables X-rays to pass through flesh, but be absorbed by bone, producing the X-ray photographs of Anna Bertha's hand—and my green twig fracture. It also explains why we try to limit the amount of X-ray radiation to which an individual is exposed; there is an increased risk of cancer from too much radiation because the high energy of the X-rays can damage cells.

The Discovery of Radioactivity

At almost the same time that Roentgen was using photographic plates to show the effects of X-rays, the French physicist Henri Becquerel was also using photographic plates to investigate the ability of materials to produce phosphorescence when exposed to sunlight. Phosphorescence differs from fluorescence in that fluorescence refers to the immediate reemission of light at a different wavelength from the electromagnetic radiation that induced it, whereas a phosphorescent material does not immediately reemit light. Becquerel had at his disposal a wide variety of substances; one of these was potassium uranyl sulfate, a uranium salt. Becquerel discovered that this material phosphoresced when exposed to sunlight—but so did several other materials. However, one day something genuinely startling occurred. Becquerel developed photographic plates that had been placed near the uranium salt but had been left in a dark drawer because there had been little sunlight the preceding days. Becquerel had expected only feeble traces of phosphorescence, but instead the outline of the crystal of uranium salt showed up sharply on the photographic plates, without being exposed to sunlight. It takes energy to produce phosphorescence. Clearly, the uranium salt itself was emanating the energy—but by what process it was producing the energy was totally unknown.

An intensive study of this process was undertaken by the husband-and-wife team of Pierre and Marie Curie, who, along with Becquerel, would jointly win the Nobel Prize for Physics in 1903 for their discoveries. Pierre Curie died on a busy Paris street on a rainy day, when he slipped on the pavement and was run over by a horse-drawn vehicle. Marie Curie continued the work and discovered radium, a material that produced energy much more vigorously than uranium. For this discovery, she was awarded the Nobel Prize in Chemistry in 1911. (She is one of only four scientists who have been awarded two Nobel Prizes, and she is the only one who has received Nobel Prizes in two distinct physical sciences. John Bardeen received two Nobel Prizes for Physics, Frederick Sanger received two for Chemistry, and Linus Pauling received both the Nobel Prize for Chemistry and the Nobel Peace Prize.[10])

Marie Curie named the phenomenon that Becquerel had discovered radioactivity. She named it, studied it—and probably died from its effects, as the illnesses that plagued her shortly after her receipt of the 1911 Nobel Prize and to which she eventually succumbed are now known either to be caused or exacerbated by prolonged exposure to radioactivity.

The Discovery of the Electron

Roentgen had been experimenting with Crookes tubes because these items had been used to study other types of rays (recall that Roentgen's paper on X-rays was titled "On a New Kind of Rays"). The fact that these rays were produced in near-vacuum conditions, when the atoms themselves were separated by distances much more vast than the separation distance in liquids and solids, prompted the speculation that individual atoms might have structure that was difficult to observe when the atoms were close together.

In 1838, Michael Faraday had discovered that when electricity was sent through a tube containing air at very low density, a glowing arc appeared from just in front of the cathode all the way down to the anode. There was a very short dark space in front of the cathode—ever the careful experimenter, Faraday noticed this—which came to be known as the Faraday dark spot. As better and better vacuums were created, the dark spot lengthened, and with the advent of Crookes tubes, the glow completely disappeared, and luminescence was seen at the opposite end of the tube. This was taken as indirect evidence that something carrying an electric charge was traveling through the Crookes tube.[11]

In 1897, the English physicist J. J. Thomson produced the first convincing evidence that there was a particle considerably smaller than a hydrogen atom that was able to carry an electric charge. His experiment was designed to measure the charge-to-mass ratio of such a particle, if it existed. Faraday had conducted experiments involving electrolysis—passing a direct current through a chemical compound in order to split it into its constituent elements—and had found that the charge-to-mass ratio of electrified hydrogen (what we now call a hydrogen ion) was about 9.65×10^7 coulombs per kilogram. Thomson's experiment was ingenious and involved little more than algebraic manipulations of known laws, so we can retrace his experiment both descriptively and algebraically.

Thomson's experiment consisted of two steps. In the first step, the particle that we now know as an electron was fired from a cathode in a Crookes tube. It was subjected to two opposing forces: a vertical electric force created between an upper positively charged plate and a lower negatively charged plate, and a horizontal magnetic force, which was known by Oersted's experiment to induce a vertical electric force. The electric force tended to deflect the electron vertically, but when the electric and magnetic forces had the same magnitude but opposite directions, there was no net deflection. Assuming the particle carried a charge e (not to be confused with the base of the natural logarithms) and the electric field strength was E, the electric force on the particle was eE. The particle traveled horizontally with a constant velocity v; from Maxwell's equations it was known that if the magnetic field strength was H, the magnetic force on the particle was evH. When the two forces balanced so that the particle continued to travel in a straight line, $eE = evH$. This equation could be solved for v, yielding $v = E / H$ as the constant horizontal velocity of the particle.

The magnetic force was created by electromagnets, which were now switched off. This didn't affect the horizontal velocity of the electron, but since there was now no force to counterbalance the electric force, the electron was subjected to the constant force eE. (The gravitational force, of course, was also acting on the particle to pull it down, but the magnitude of this force was so small in comparison with the electric force that we can ignore it here.) By Newton's second law, the force $eE = ma$, where m was the mass of the particle and a its acceleration. The particle traveled a distance L horizontally and d vertically before striking the positively charged plate. Since the horizontal velocity was $v = E / H$, the time T that it took to travel the horizontal distance L was given by $L = vT$, so $T = L / v = LH / E$. The movement of the particle in the vertical direction under the constant acceleration a was analogous to the movement of a falling object under gravitational acceleration; the well-known formula for the distance s that an object falls in time t is given by $s = \frac{1}{2}gt^2$, where g is the acceleration of gravity at the surface of the Earth, which we saw in Chapter 1. In this case, the distance d that the electron moved vertically was given by $d = \frac{1}{2} aT^2$, because it moved vertically a distance d in the same time that it moved horizontally a distance L.

It's now merely a matter of algebra to find the charge-to-mass ratio of the electron. From the equation $eE = ma$, we see that the charge-to-mass ratio is $e / m = a / E$. From $d = \frac{1}{2} aT^2$ we see that $a = 2d / T^2 = 2dE^2 / L^2H^2$, and so finally $e / m = 2dE / L^2H^2$, where all the quantities on the right could be measured quite accurately. On doing the computations, Thomson found that the charge-to-mass ratio was 1,000 times greater than the charge-to-mass ratio of the hydrogen ion. It also didn't matter what gas he used to create these particles; the results were the same.

There were several possible conclusions, which Thomson outlined in his write-up of the experiment. He worked with the mass-to-charge ratio m / e rather than the charge-to-mass ratio e / m we derived above, but of course the mass-to-charge ratio is the reciprocal of the charge-to-mass ratio. Thomson wrote, "The smallness of m / e may be due to the smallness of m or the largeness of e, or to a combination of these two." However, the fact that cathode rays could move unimpeded through dense collections of atoms persuaded Thomson that the particles were smaller in size than ordinary atoms. Two years later, Thomson demonstrated the same charge-to-mass ratio for particles freed when metals were irradiated by ultraviolet light. The irradiated metals acquired a positive charge because the ultraviolet light imparted sufficient energy to free electrons; this was the photoelectric effect that Einstein explained during his "miracle year."[12]

$E = mc^2$

Einstein's famous equation of energy and mass is probably the iconic intellectual achievement of mankind. I've never actually seen any survey on its Q score (a measure of recognition), but my guess is that almost every high school graduate has at least seen $E = mc^2$ and realizes that it has something to do with energy, Einstein, or both. This has become recognized as the supreme achievement of Einstein's "miracle year" and, together with the discovery of subatomic particles, pointed the way for the explanation of *how* the Sun keeps on shining.

The equation says simply that if an object with mass m is converted completely to energy, the energy E that is obtained is the product of that mass times the square of the speed of light. In order to appreciate the immense amount of energy stored up in even a small amount of mass, a little calculation is in order.

The joule, the basic unit of energy in the modern form of the metric system, can be defined as the energy required to accelerate a 1 kilogram mass to a speed of 1 meter per second, over a distance of 1 meter, and in 1 second—that is, 1 joule = 1 kg × m^2 / s^2. It takes more than 4,000 joules to heat a kilogram of water by 1 kelvin (or degree Celsius), so a joule is, by itself, not very much energy. The amount of energy released by the first atomic bomb is on the order of 80 terajoules, or 8×10^{13} joules. Amazingly, as Einstein's energy-mass equation demonstrates, a single dollar bill, under the right circumstances, could be an equally dangerous weapon. The speed of light is 300,000 kilometers per second, so $c = 3 \times 10^8$ meters per second, and a dollar bill has a mass of about 1 gram, or .001 kilograms. Therefore, if a dollar bill were completely converted to energy, it would yield a total of $.001 \times (3 \times 10^8)^2 = 9 \times 10^{13}$ joules, a little more than the energy released by the detonation of the first atomic bomb.

More than a decade after the publication of $E = mc^2$ in what was to become known as the special theory of relativity, Einstein published the general theory of relativity, which extended his special theory to encompass Newton's theory of gravitation. A key difference between Newton's and Einstein's theories was the bending of light near a heavy object such as the Sun; the 1919 total eclipse provided the opportunity for scientists to determine whether Einstein's general theory was entitled to supplant Newton's. The expedition that performed the measurements confirming the general theory was led by the renowned British astrophysicist Sir Arthur Eddington.[13]

In the following year, Eddington gave an address to the British Association for the Advancement of Science. His subject was *how* the Sun, and other stars, keep on shining. Eddington observed that "a star is drawing on some vast reserve of energy by means unknown to us. This reservoir can scarcely be other than the subatomic energy which, it is known, exists abundantly in all matter; we sometimes dream that man will one day learn how to release it and use it for his service."[14] He concluded his lecture with the prescient declaration that "if, indeed, the subatomic energy is being freely used to maintain their great furnaces, it seems to bring a little nearer to fulfillment our dream of controlling this latent power for the well-being of the human race—or for its suicide."[15]

How the Sun Keeps on Shining

Although Thomson had provided strong evidence for the existence of negatively charged particles within the atom, atoms by themselves are electrically neutral. Consequently, there must have been positively charged particles lurking within the atom to neutralize the negatively charged electrons. Although it was originally thought that the positively and negatively charged particles were distributed uniformly throughout the atom, much as raisins and blueberries are uniformly distributed throughout a blueberry-raisin muffin, experiments by Ernest Rutherford showed that these positively charged particles, called protons, were tightly clustered together in what is now called the nucleus of the atom.

In the same year that Eddington delivered his speech on the subatomic energy that was needed to power the Sun, Rutherford conjectured that it might be possible for an electron and a proton to combine to form a particle that was electrically neutral. Twelve years later, the neutron, a particle with no electrical charge, was discovered. This particle helped to explain the structure of the helium atom, which was known to have an atomic number of 2 but an atomic weight of 4; its nucleus had to contain two protons, in order to have an atomic number of 2, and two neutrons, to bring its atomic weight up to 4.

At the time that Eddington gave this speech, it was known that the two primary constituents of the Sun were hydrogen, which accounted for 71 percent of the mass of the Sun, and helium, which accounted for 27 percent of its mass. Eddington had conjectured that a possible source of the subatomic energy he had discussed might be obtained by slamming together hydrogen nuclei.

There were a number of theoretical obstacles to be overcome. The atomic weight of four hydrogen atoms was 4—the same atomic weight as an atom of helium—but the four hydrogen atoms contained four protons and the helium atom contained two protons and two neutrons. There needed to be a way of getting the four protons to morph into two protons and two neutrons. Additionally, the probability of the simultaneous collision of four atoms seemed unlikely—much as the simultaneous collision of four automobiles is unlikely. But although four-car *collisions* almost never occur, four-car *pileups* occur with some frequency, as two cars collide and other cars cannot avoid the obstacle and smash into the existing pile.

Another problem was the electric repulsion of the protons, whose strength we know from a previous chapter to be greater than the gravitational force between them by many orders of magnitude. However, it was possible that under high heat (which is equivalent to large velocity), the protons would have enough energy to overcome the electrical repulsion between them. The problem was that the temperature of the Sun seemed to be insufficient to make this happen.

The solution was found in an exquisite synthesis of ideas from quantum mechanics and statistical mechanics, combined with a flood of new experimental data. Although the average temperature of the Sun was not high enough to allow two protons to collide, statistical mechanics provided a distribution for the temperatures of the molecules that showed that an infinitesimal fraction had temperatures sufficiently high enough to undergo a peculiar procedure known as quantum tunneling. This process enabled the protons to slam together and shed one of the electrical charges as one of the protons was transformed into a neutron. Analogous to the four-car pileup discussed previously, there is a sequence of nuclear reactions, known as the proton-proton chain, that produces the nucleus of a helium atom from the nuclei of four hydrogen atoms.[16]

In so doing, a small fraction of the mass of the hydrogen atoms is converted into energy. The atomic weight of hydrogen is about

1.00794, and the atomic weight of helium is 4.0026. The weight of an electron is about .00055, so the atomic weight of the nucleus of a hydrogen atom is $1.00794 - .00055 = 1.00739$, and the atomic weight of the nucleus of a helium atom is $4.0026 - 2 \times .00055 = 4.0015$. The mass lost when four hydrogen nuclei become one helium nucleus is therefore $4 \times 1.00739 - 4.0015 = .02806$, and the weight loss fraction per nuclei is $.02806 / (4 \times 1.00739) =$ approximately .007.

The proton-proton chain is not the only means by which fusion is produced, although it accounts for most of the thermonuclear fusion that takes place in the Sun. There is an alternate procedure known as the CNO cycle, in which the elements carbon, nitrogen, and oxygen are both produced and act as midwives to produce helium from hydrogen. This process occurs at higher temperatures than the proton-proton chain, and so occurs more rarely in the Sun, but becomes more important in hotter stars.[17]

Thunderball

I sometimes speculate that Ian Fleming, the creator of James Bond, had somehow learned of the efficiency of hydrogen fusion. He did, after all, attend Eton College, the premiere private school in England—although it seems unlikely this topic would arise at the prep school level. Nonetheless, James Bond wasn't known as "7," he was known as "007". Maybe he went with it, not because of anything to do with the efficiency of hydrogen fusion, but because it was just right.

Well, physically speaking, .007 had to be just right, too—it's a real Goldilocks number, and perhaps the most idiosyncratic one we've looked at in this book. The other numbers are universal constants. As far as we know, the gravitational constant G is the same here as it is in galaxies ten billion light-years distant. Avogadro's number is the number of molecules in a mole of any substance—although the idea of a mole is defined such that Avogadro's number cannot vary from substance to substance. But it probably wouldn't matter much if those numbers were a bit different. But .007, on the other hand, can't change without things becoming very different. Thermonuclear fusion affects other transformations in addition to creating helium from hydrogen, although the efficiency of these is considerably less than the efficiency of creating helium from hydrogen. This is a very good thing—for us. If the efficiency of these transformations were too high, the lives of stars would be much shorter, and there probably wouldn't be time for life to evolve.

The steps in thermonuclear fusion are exquisitely sensitive to the number .007. The first step in the proton-proton chain, when two protons slam into each other, produces deuterium, an isotope of hydrogen whose nucleus contains one proton and one neutron, because one of the protons "sheds" its electrical charge and morphs into a neutron. If the efficiency were as low as .006, the neutron and proton would not bond to each other, deuterium would not form, and the universe would consist of nothing but hydrogen. There would still be stars, but instead of being engines of creation for other elements, they would simply be large and sterile balls of hydrogen that—as in Kelvin's second attempt at determining how the Sun keeps on shining—heat as they contract under gravitation, glow, and eventually cool and die.

If, on the other hand, the efficiency of hydrogen fusion were as high as .008, it would be far too easy for protons to bond together. All the hydrogen in the universe would rapidly form helium and heavier elements, and without hydrogen, there would be no water. We could speculate on other forms of life having the possibility of emerging in such a universe, but they would be vastly different from anything we know—and they certainly wouldn't be us.

CHAPTER 11

THE CHANDRASEKHAR LIMIT

In 1835, the philosopher Auguste Comte attempted to go where no philosopher had gone before. Hitherto philosophers had tried to define the limits of human knowledge, but had generally done so by looking at moral, ethical, and religious questions they felt would never be resolved. Comte compiled a list of questions he felt science would never be able to answer. One of those questions was to determine the composition of the stars. Comte's belief—that this was a question science would never answer—was not an unreasonable one. After all, in 1838 Friedrich Bessel would show that the distance to the star 61 Cygni was six light-years, a distance almost 400,000 times as long as the distance from the Earth to the Sun, and far larger than anyone had previously suspected. Finding out anything about so distant an object would appear to be immensely challenging.

Philosophers and scientists generally don't read the same books and journals, fortunately, and so Comte's prediction undoubtedly went unheeded by Robert Bunsen.[1] A German chemist, Bunsen had been absorbed in studying organic arsenic-containing compounds at the time of Comte's list. Understanding such compounds would prove to be a milestone in the history of science, but not for some seventy-five years, when Paul Ehrlich used them to develop a cure for syphilis. In Bunsen's time, such compounds were merely very dangerous: Bunsen lost an eye and almost died twice from arsenic poisoning. Upon recovering, he decided that discretion was the better part of valor and abandoned organic chemistry for a safer endeavor, a study of the role of heat in chemical reactions. Among other great successes, this would lead to the development of the Bunsen burner, a device known to every student who has ever set foot in a chemistry lab.

One of Bunsen's early students was Gustav Kirchhoff; together they collaborated on investigations of chemical reactions that absorb or emit light. The two combined Thomas Young's idea of passing light through a slit with Newton's idea of passing light through a prism. Thus was born the spectroscope, one of the most important tools in science. The Bunsen burner was used to heat material to incandescence, and the light the incandescent material emitted was passed through a spectroscope to throw a pattern of colored lines on a screen. It was soon discovered that this pattern of colored lines was a chemical fingerprint, and that each element had its own characteristic pattern, or spectrum.

The spectroscope turned out to be an extraordinary tool. Using the spectroscope to analyze the light of the sun, Kirchhoff discovered a spectral line characteristic of the element sodium. Since there was no sodium in the Earth's atmosphere, and certainly none in the vacuum between the Sun and the Earth, the conclusion was inescapable: sodium existed in the Sun. The same technique would later be used on the light from stars, enabling their chemical composition to be determined, not even thirty years after Comte's gloomy prediction. The fame Kirchhoff's work brought him was met with skepticism in at least some quarters, however. Kirchhoff's banker, a pragmatic sort, asked Kirchhoff, "Of what use is gold in the sun if I cannot bring it down to Earth?" Shortly after this remark was uttered, Kirchhoff received a prize from Great Britain for his work, awarded in golden sovereigns. Kirchhoff, handing them to his banker, seized the opportunity to remark: "Here is gold from the sun."[2]

Comte died two years before the development of the spectroscope. Comte was wrong about the composition of the stars, but he was correct in principle: there are some aspects of the universe that science will never be able to determine. Some systems, such as global weather, are so sensitive to small changes that we simply cannot predict the weather with great accuracy more than a few days in advance. The problem with statements like Comte's is that the main effect of labeling something as impossible to know is only to make it less likely that anyone will try. Better, like Albert Einstein, to assume that if we have a question, we can find an answer, and so begin asking questions."[3] If I had to guess, the vast majority of children ask most, if not all, of the following three simple questions. The first is "Where did I come from?" a question generally asked far too early for most parents by most children. (My parents, like most parents, fumbled the ball—mine didn't really supply a satisfactory answer until I was about ten or eleven, and then passed the buck by handing me an extremely boring book on the human reproductive system. Possibly they figured that if I got sufficiently bored with the subject, I wouldn't trouble them again. They were right—like many of my contemporaries, I learned the real story in dribs and drabs from a variety of sources.) The second and third questions are variations on the theme of the first question, which is the problem of origins. Children are so into their own world that the first question concerns their own individual origin. As their spheres of awareness expand, the other two questions follow naturally, the second question being "Where did the world come from?" and the third, "Where did it all come from?" Einstein was right: these simple questions are indeed profound, and have motivated many of the most important advances in scientific understanding.

Of course, a motivation is only as good as the tools in its service, and for answering big questions in science, Bunsen and Kirchhoff's spectroscope is hard to beat. If I had to build a top-ten list of scientific tools, I'd be hard-pressed not to put it at the top, although the microscope, with its great impact on human health, has much to recommend it. Top-ten lists have a lot going for them, too: I might not have any interest in architecture (I don't), but if I happen to come across a list on the web of the top-ten most important (or largest, or most beautiful) architectural structures, I'll probably click on it. In fact, there is one top-ten list that is of immense importance in answering my second and third questions above, and that is the list of the ten most common elements in the solar system. We have such a list thanks to the spectroscope, which these days enables us not just to determine what's in the Sun, or the solar system, or the universe, but how much of it there is, too.

ELEMENT	% OF ATOMS
Hydrogen	92.295
Helium	7.548
Oxygen	0.082
Carbon	0.048
Nitrogen	0.009
Neon	0.008
Magnesium	0.003
Silicon	0.003
Iron	0.002
Sulfur	0.002

What the list makes clear is how unlikely our home planet is. In fact, any concentrated assemblage of stuff is extremely unlikely; after all, the universe itself is incredibly empty, with an average density of only one atom for every five cubic meters of space. Of course, gravity and electromagnetism help that stuff stick together, but what makes Earth seem even unlikelier is that it's not just an assemblage of hydrogen and helium with a bit of other stuff thrown in. Instead, where the universe has hardly any heavy elements in it, our planet has quite a lot, with oxygen, aluminum, silicon, sodium, potassium, calcium, and iron all being at least as common on Earth as hydrogen is.

One can see just how different the distributions of different types of matter are on our planet and in the universe at large. The human body is about 65 percent oxygen and 19 percent carbon by weight. I weigh 140 pounds, so my body contains about 91 pounds of oxygen and 26 pounds of carbon (more or less). A mole of oxygen weighs 16 grams, and since there are 454 grams in a pound, my body contains 2,582 moles of oxygen, or 1.56×10^{27} atoms of oxygen. If the matter in the universe were distributed evenly, there would be only one oxygen atom per 6,250 cubic meters of space, which means you would need 9.75×10^{30} cubic meters to supply the oxygen in my body. That's a cube 2.14×10^{10} meters—13 million miles—on a side, more than fifty-seven times the distance of the Earth to the Moon.

Okay, so the universe went to a lot of trouble to concentrate enough carbon and oxygen on Earth to make this book possible: supplying carbon and oxygen to make author, readers, and paper available. Unearthing the story of how this happened started with the invention of the spectroscope, and took almost a century to reach the conclusion so memorably phrased by Carl Sagan, "We are made of star-stuff."[4] Behind that conclusion lies something known as the Chandrasekhar limit.

Balancing Act

Chemical reactions involve the breaking and re-forming of chemical bonds; the elements involved rearrange themselves into new chemical compounds. Keeping track of this is done by means of a bookkeeping system called "balancing equations."

A simple example of a chemical reaction is one that takes place when sodium hydroxide, commonly known as lye, is mixed with sulfuric acid. (Stand clear if you do it; the reaction is quite violent.) Sodium hydroxide has the chemical formula NaOH; sulfuric acid has the formula H_2SO_4 (when I learned chemistry, we learned the rhyme "Willie studied chemistry, he studies it no more, for what he thought was *H-two-O*, was *H-two-S-O-four*"). This reaction is written

$$2NaOH + H_2SO_4 \rightarrow Na_2SO_4 + 2H_2O$$

This is chemical shorthand for two molecules of sodium hydroxide reacting with one molecule of sulfuric acid, yielding one molecule of sodium sulfate (Na_2SO_4) and two molecules of water. The arrow tells us which way the reaction proceeds (the stuff that reacts is on the tail side of the arrow, the stuff that you end up with is on the pointed side of the arrow). The reaction is balanced because the totals are the same on each side of the arrow for all the elements involved. There is one atom of sulfur on each side of the equation, two atoms of sodium, four atoms of hydrogen, and six atoms of oxygen.

However, this description is incomplete. Just as in the real world, some aspects of transactions are "off the books," in this case, a key player, energy, has not been mentioned. Some reactions are endothermic; they require energy in order to generate the reaction. One example of this that we have encountered so far in this book is the electrolysis of compounds into constituent atoms. For example, in order to separate water into its component elements (hydrogen and oxygen), it is necessary to supply energy in the form of electricity. The electrolysis of water would be written

$$2H_2O + \text{energy} \rightarrow 2H_2 + O_2$$

To do a really thorough job of bookkeeping, we should specify the quantity of energy involved, but we won't go into that level of detail here.

Other reactions are exothermic; the reaction produces energy. A terrific example of this occurs inside your automobile engine when ethane (C_2H_6) is combined with oxygen at a high enough temperature to ignite the ethane. The reaction produces water and carbon dioxide—and I'm sure you're familiar with the role carbon dioxide may play in producing global warming. When we include energy, the reaction is written

$$2C_2H_6 + 7O_2 \rightarrow 4CO_2 + 6H_2O + \text{energy}$$

The essence of all of this is that chemical reactions conserve the number of atoms of each element. This is the reason that the alchemists' search for the philosopher's stone, whose touch would transmute base metals to gold, was doomed to failure—alchemists had only chemical reactions at hand. In order to produce nuclear reactions, which can change the kinds of atoms present, you need either much more impressive technology than alchemists had or a lot of heat. The former only became available in the twentieth century, and the latter can only be found in the heart of a star.

Nucleosynthesis

The answer to a child's first deep question—"Where do I come from?"—usually begins with some hemming and hawing about what went on nine months before the child's birth, but without the element carbon the whole issue would be moot. Carbon, like most elements other than hydrogen and helium (which were created shortly after the big bang), is the result of nucleosynthesis, a process of element creation that occurs most frequently in stars. The particular reaction that produces carbon is known as the triple-alpha process, and is analogous to not just a nearly simultaneous three-car pileup, but one of identical cars—say, three blue 2006 Toyota Corollas. In balancing nuclear reactions, one does not balance the books element by element, as is done in chemical reactions, but by atomic number (which is the number of protons in the nucleus) and the total number of protons and neutrons in the nucleus. The element helium-4, which has a nucleus of two protons and two neutrons, is denoted by ^4_2He. The triple-alpha process consists of an endothermic reaction, in which two helium atoms fuse to create a beryllium atom, followed by an exothermic one in which the beryllium atom fuses with a helium atom to create a carbon atom. The two equations are

$$^4_2\text{He} + {}^4_2\text{He} \rightarrow {}^8_4\text{Be}$$
$$^4_2\text{He} + {}^8_4\text{Be} \rightarrow {}^{12}_6\text{C} + e^+ + e^-$$

In addition to the carbon atom, the second reaction also produces a positron (the second term on the right) and an electron (the last term). The two reactions also show a net production of approximately 1.16×10^{-12} joules. That's not much—and the triple-alpha process is considerably less likely than a three-car pileup of Toyota Corollas (even though these weren't the models affected by the problem with the stuck accelerator, which certainly would increase the probability of such a collision). Making it even more difficult is the fact that it requires temperatures in excess of 100 million degrees kelvin to get the ball rolling, and these temperatures are available only at the center of *really* big stars. Nonetheless, there have been enough such stars in the past that carbon is the fourth most abundant element in the universe.

Both the remarkably high concentration of oxygen and carbon, and the general likelihood of unusual three-car nuclear pileups, raises another simple, but very big, question. It's one of the great recurrent questions in science and philosophy, and *not* one liable to occur to children: namely why the universe seems so precisely arranged as to enable the evolution of *Homo sapiens*. Indeed, we've already seen another example of such a delicate arrangement—the efficiency of hydrogen fusion. Prior to 1952, the triple-alpha process was not known to exist or even to be possible, but the astrophysicist Fred Hoyle was able to argue compellingly—through a sort of intellectual reverse-engineering—that not only must it be possible, but that it must exist. We exist, he argued, and to exist as we do, we rely on ample carbon. Therefore, any process of nucleosynthesis must include a means to create it, and that requires the triple-alpha process, no matter how unlikely it seems that it would actually happen. QED.

But there is more to the story. Just as the efficiency of hydrogen fusion is nicely arranged to enable the development of *Homo sapiens* (or at least life on Earth), so is the carbon atom. The chemical properties of the carbon atom enable the biochemistry of our kind of life, and the nuclear properties of the carbon atom enable carbon to exist in enough abundance that life can get started. Those nuclear properties not only enable the carbon-producing triple-alpha process to take place, but they also prevent a "quadruple-alpha" process, in which carbon fuses with helium to produce oxygen, from occurring with enough frequency to burn up the carbon. Yes, we need oxygen, but we need carbon first— life forms evolved on Earth long before oxygen was present in the atmosphere in significant amounts. If the reaction that produces oxygen from carbon were more common, there would be more oxygen to breathe, but nothing around to breathe it.

The discovery of the list of reactions that constitute the nucleosynthesis story is one of the great achievements of twentieth-century science, although it doesn't usually get the publicity. What has evaded us so far is the ability to reproduce nucleosynthesis here on Earth. Even the fusion of hydrogen to helium, the simplest of the nucleosynthesis reactions, requires a tremendous amount of energy. We have been able to duplicate it explosively in the detonation of a hydrogen bomb, but only if we first set off an atomic bomb to produce the necessary temperatures and pressures. It's hardly the sort of thing that calls out for industrial applications. We have achieved fusion under controlled conditions in laboratories, but have not yet managed to do so in a cost-effective fashion. If we can accomplish that, we will have a clean source of energy that will probably supply humanity's energy needs for millennia. But the fusion reactions needed to create the vast complex of heavier elements are far beyond our abilities to produce—they require such incredible temperatures and pressures that the only place they can be manufactured is in truly massive stars.

Why Does the Sun Go On Shining?

On further thought, it seems maybe the country-western songs *can* ask really deep questions—if we cut them a little slack. We answered the question of *how* the Sun shines when we discussed the process of nuclear fusion, but why does it go *on* shining? Why does it keep shining—and shining—and shining? Part of the answer lies in the efficiency of hydrogen fusion; there's an awful lot of hydrogen in the Sun, and it doesn't fuse very efficiently, so it will take a long while to use it up. However, the fact that there's an awful lot of hydrogen in the Sun means that the gravitational force exerted by the hydrogen is very strong, and since gravitational force is attractive, it is acting to pull all the hydrogen to the center of the Sun. Why doesn't the Sun collapse?

We know that as gas gets hotter, it expands; and it is the balance between the outward thermal pressure from hydrogen fusion and the inward compression from gravitation that keeps the Sun in equilibrium—over periods of years, decades, centuries, and millennia, even over millions of years. But not over more than a few eons (an eon is a billion years). Gravitational compression is relentless, and in order for a star to keep shining, it must find a way to keep exerting outward pressure to balance that compression. So, when a star fuses its hydrogen to helium, is that all she wrote?

As we have seen, there are other fusion processes available; otherwise carbon would never be produced. However, it's a lot more difficult to fuse helium than it is to fuse hydrogen. In order to fuse hydrogen, one needs to bring hydrogen atoms sufficiently close together to overcome the Coulomb barrier, the repulsion generated when the electron in one hydrogen atom is brought near the electron in another hydrogen atom. We saw in a previous chapter that the force of the electric repulsion between two electrons was more than 10^{39} times stronger than the force of the gravitational attraction between them. To overcome this repulsion requires extremely high temperatures; the actual process by which electrons overcome the Coulomb barrier is not through the simple application of slamming them together at really high speed, but through a more subtle quantum mechanical process known as quantum tunneling. Quantum tunneling takes place because electrons are not really fast-moving dots as they are conventionally pictured. In fact, a good argument could be made that nobody really knows exactly what an electron is; the best description we have of them for computational purposes is as a mathematical construct known as a probability wave. Electrons do not really have a definite position in space like everything in the macroscopic world. Whatever electrons are, we can say where they are most likely to be, but the fact is that they can be anywhere—and the higher the temperature, the more likely they are to be somewhere else—like on the other side of the Coulomb barrier.

Heavier atoms have more electrons, and so the electrical repulsion between heavier atoms is greater than the electrical repulsion between hydrogen atoms. This means that even higher temperatures are required to make the atoms move with enough speed for their electrons to be able to tunnel through the Coulomb barrier. The only way to get those temperatures is with greater compression from the gravitational force—but this will tend to happen because every time two hydrogen atoms fuse to a helium atom, the total number of atoms decreases by one. When all the hydrogen has fused to helium, only a very small fraction of the mass has been converted to energy—the iconic .007 that represents the efficiency of hydrogen fusion—but there are only half as many atoms. Gravitational compression acts to confine these atoms to a smaller space—which heats the atoms. If the star were large enough to begin with, there would be enough helium to enable gravitational compression to elevate temperatures to the point where helium fusion can begin.

And the story repeats. After all the helium has fused, there is almost as much mass as there was at the start of helium fusion, but a lot fewer atoms. Gravitational compression forces these to occupy an even smaller volume, heating the star still further, and under the right conditions, enabling fusion of even heavier atoms.

However, the rapidity of this process does not scale linearly. Helium fusion to carbon proceeds much more rapidly than hydrogen fusion to helium. This explains why it took so long for life to evolve, because it takes a long time for hydrogen to fuse to helium in order to set the stage for the fusion of helium to the carbon that will enable the creation of life. It also explains why life has the opportunity to evolve: because once there is a planet with lots of carbon circling a star like the Sun, the efficiency of hydrogen fusion to helium enables that star to be stable for eons.

In fact, the life cycle of a really heavy star, one with a mass of twenty times that of the Sun, is like a play with shorter and shorter acts of ever-increasing dramatic tension. In such a star, it takes roughly a billion years for the hydrogen to fuse to helium, but only about a million years for the helium to fuse to carbon and oxygen. It takes perhaps 100,000 years for the carbon to fuse to neon and magnesium. The oxygen burns to silicon and sulfur in twenty years, and the silicon and sulfur burn to iron in a week! The different rates at which these processes take place leave the star looking like a multilayered Tootsie Pop: a heavy iron core overlaid by a spherical shell of silicon and sulfur. As we proceed toward the surface of the star, we encounter a succession of cooler spherical shells: neon and magnesium, then carbon and oxygen, then helium, and then, at the outside, hydrogen.

What happens next turned out to be a fascinating story, deciphered by a fascinating individual—Subrahmanyan Chandrasekhar, more familiarly known to his friends and colleagues as Chandra.

Chandra

Some professors, like Isaac Newton, frankly suck, and find themselves continually lecturing to empty or near-empty rooms. Some professors, like Ludwig Boltzmann, are inspiring and beloved by their students. Some professors inspire a mystique; they appear brilliant but unapproachable. I had such a professor, Shizuo Kakutani, who taught the mathematical analysis course at Yale, and I wouldn't be surprised if practically everyone who has ever attended college can recall such a professor, no matter whether they studied mathematics, history, or literature. Chandrasekhar was evidently such a teacher, as no less a student than Carl Sagan would later recall.

> Chandra was giving a colloquium. Three walls of the lecture room had blackboards on them, all spotlessly clean when Chandra began his lecture. During the course of his lecture, he filled all the blackboards with equations, neatly written in his fine hand, the important ones boxed and numbered as though they had been written in a paper for publication. As his lecture came to an end, Chandra leaned against a table, facing the audience. When the chairman invited questions, someone in the audience said, "Professor Chandrasekhar, I believe, on blackboard . . . let's see . . . 8, line 11, I believe you've made an error in sign." Chandra was absolutely impassive, without comment, and did not even turn around to look at the equation in question. After a few moments of embarrassing silence, the chairman said, "Professor Chandrasekhar, do you have an answer to this question?" Chandra responded, "It was not a question, it was a statement, and it is mistaken," without turning around.[5]

Such a story might give the impression of a cold, aloof scholar who did not care about his students. Those individuals, too, exist in practically every department, especially in top-rank institutions—but Chandrasekhar was not one of them. For much of his career, Chandrasekhar lived in Williams Bay, Wisconsin—near the Yerkes Observatory—and traveled to Chicago to teach his class, which at one point was just two students. This was in the era before the interstates, and was a journey of approximately 100 miles, but the drive does seem to have been worth it: when the Nobel Prize for Physics was awarded in 1957,[6] it went to Chen Ning Yang and Tsung-Dao Lee—the two students in that class.

Chandrasekhar, too, would win the Nobel Prize, but not until 1983.[7] It was a long time to wait—the work he won it for began when he was in his teens! By the 1920s, astrophysicists had been able to work out that a dim white star known as Sirius B, a companion star to the famous Dog Star Sirius, has an astoundingly high density, more than a million times the density of the Sun. This puzzled astronomers of the time because it was simply impossible for atoms to be squeezed to that density and retain their identity as atoms. Such densities can only be achieved if the atoms are squished so much that the electrons are no longer bound to the nucleus, so that what had been a star made of atoms becomes a star made of positively charged ions surrounded by a densely packed sea of electrons. As we saw in an earlier chapter, when the electrons are so close to one another, quantum mechanics dictates that they exert a special type of force known as electron degeneracy pressure. The Pauli exclusionary principle describes how no two particles can have precisely the same quantum state; in this case, that means that some of the electrons in the sea are forced into very high energy states, and so have extremely high velocities. This energy supports the star against the crushing force of gravity. A typical white dwarf has a mass that is roughly the mass of the Sun, but that mass is squeezed into a volume the size of the Earth.

The discovery that electron degeneracy pressure could enable a white dwarf to have hitherto unheard-of densities was made by Ralph Fowler in 1926.[8] It occurred to Chandrasekhar, a student of almost inconceivable brilliance, who was reading research-level papers before he was eighteen years old, that Fowler's paper had not taken into account the relativistic effects that would occur when the electrons were moving at extremely high velocities. What Chandrasekhar discovered when he applied his relativistic approach to Fowler's work was no mere correction: it was breathtakingly unexpected. Chandrasekhar had found a firm upper limit for the mass of a white dwarf, or any body of electron degenerate matter.

Chandrasekhar, by then Fowler's doctoral student, described his findings in a paper titled "The Maximum Mass of Ideal White Dwarfs."[9] The maximum mass depends on several of the universal constants that have already been discussed in this book: the gravitational constant, the speed of light, and Planck's constant, as well as on the number of nucleons (protons and neutrons) per electron in the star. The accepted modern value of the Chandrasekhar limit is approximately 1.4 times the mass of the Sun.[10]

This result was obtained while he was traveling on a steamship from India to England—and before he was twenty years old![11] It brought Chandrasekhar into conflict with Sir Arthur Eddington, one of the pre-eminent astrophysicists of his era. The argument was to have a profound effect on Chandrasekhar's career. At the time, one of the great problems of astrophysics was to determine the life cycle of stars. Eddington, who had devoted a large portion of his career to the problem, believed that the white dwarf stage was the eventual fate of every star, no matter how large. The conflict came to a head at a meeting of the Royal Astronomical Society in January 1935. Both Chandrasekhar and Eddington had submitted papers, but Eddington also was invited to say a few words. They were devastating:

> Fowler used the ordinary formulae [to solve the problem]; Chandrasekhar, using the relativistic formula which has been accepted for the last five years, shows that a star of mass greater than a certain limit M remains a perfect gas and can never cool down. The star has to go on radiating and radiating, and contracting and contracting until, I suppose, it gets to a few kilometers radius, when gravity becomes strong enough to hold in the radiation, and the star can at last find peace.
>
>   Dr. Chandrasekhar had got this result before, but he has rubbed it in, in his last paper; and when discussing it with him, I felt driven to the conclusion that this was almost a reduction ad absurdum of the relativistic degeneracy formula. Various accidents may

intervene to save a star, but I want more protection than that. I think there should be a law of Nature to prevent a star from behaving in this absurd way!

If one takes the mathematical derivation of the relativistic degeneracy formula as given in astronomical papers, no fault is to be found [here Eddington tosses Chandrasekhar a bone]. One has to look deeper into its physical foundations, and these are not above suspicion. The formula is based on a combination of relativity mechanics and nonrelativity quantum theory, and I do not regard the offspring of such a union as born in lawful wedlock. I feel satisfied that the current formula is based on a partial relativity theory, and that if the theory is made complete the relativity corrections are compensated, so that we come back to the "ordinary" formula.[12]

While Eddington did not challenge the accuracy of Chandrasekhar's derivations, he implied that Chandrasekhar had made a fundamental error in the underlying physics in order to reach such an apparently absurd conclusion. Chandrasekhar returned from the meeting utterly depressed. After all, Eddington was an established giant in the field, but Chandrasekhar did not give up. He began to communicate with many of the eminent physicists of the day in an attempt to determine whether he or Eddington had analyzed the situation correctly from a physical standpoint. The weight of opinion was squarely on Chandrasekhar's side. As the eminent physicist Rudolf Peierls recalled, "I did not know any physicist to whom it was not obvious that Chandrasekhar was right in using relativistic Fermi-Dirac statistics, and who was not shocked by Eddington's denial of the obvious, particularly coming from the author [Eddington] of a well-known text on relativity. It was therefore not a question of studying the problem, but of countering Eddington."[13]

For Chandrasekhar, countering Eddington did not mean war. Throughout the battle, he and Eddington had maintained warm personal relations—this was not the stuff of soap opera. In recalling this period, Chandrasekhar remarked, "It never destroyed my respect for him. . . . It never gave me a feeling that I was not on speaking terms with him. . . . During the spring of that year (right after the meeting of the Royal Astronomical Society), we went on a bicycle trip together and Eddington took me to the Wimbledon tennis championship match."[14] In fact, that friendliness put Chandrasekhar in something of a bind. He could have gone for the kill—the support of the world's top physicists was all the vindication of his work he needed. Instead, he left Eddington alone, and set aside his research on white dwarves to work on other problems, and to avoid embarrassing his friend. When Eddington died in 1944, Chandrasekhar paid Eddington the following tribute:

> I believe that anyone who has known Eddington will agree that he was a man of the highest integrity and character. I do not believe, for example, that he ever thought harshly of anyone. That was why it was so easy to disagree with him on scientific matters. You can always be certain he would never misjudge you or think ill of you on that account.[15]

Chandrasekhar didn't abandon the work forever. When pulsars were discovered in the 1960s, Chandrasekhar returned to the study of stellar structures in an effort to explain their workings, continuing the work he had begun nearly three decades before. In 1983, Chandrasekhar was one of two physicists to share the Nobel Prize for Physics; even though his career had included monumental contributions to a wide variety of topics in astrophysics, the Prize was essentially awarded for the work he had done while on a steamer in the summer of 1930. Chandrasekhar himself disarmingly summarized his life by saying, "I left India and went to England in 1930. I returned to India in 1936 and married a girl who had been waiting for six years, came to Chicago, and lived happily thereafter."[16]

Mozart was writing symphonies when he was five. Olympic gold medals have been won by twelve-year-olds, and Alexander the Great conquered the world by the time he was twenty. All of these are amazing accomplishments, yet I am more awed by the ability of a student with only two years of college behind him to assimilate the most brilliant theories of his day, and use them to decipher the secrets of the stars. The physicist Res Jost said it best: "There is a secret society whose activities transcend all limits in space and time, and Dr. Chandrasekhar is one of its members. It is the ideal community of geniuses who weave and compose the fabric of our culture."[17]

The upshot to all of this is what happens when a star has a mass greater than the Chandrasekhar limit. Rather than becoming a white dwarf, it explodes in a supernova, casting all those heavy elements— everything up to iron—out into the Universe. In fact, the explosion is so energetic that even more fusion takes place, creating those elements beyond iron that fusion pathways in a star cannot. A little radioactive decay later on down the road gives us the lighter ones to round out the periodic table.

Every airplane has a take-off speed: the speed necessary for it to safely become airborne. The Chandrasekhar limit is not just a number that tolls the death knell for a massive star, it is the take-off speed for the formation of planets—and life.

CHAPTER 12

THE HUBBLE CONSTANT

———

In 1758, Charles Messier, a French astronomer, trained his telescope on the heavens, hoping to be the first to record the return of the great comet of 1682, which had been predicted by the English astronomer Edmond Halley. It was not to be, as Messier wasn't the first to spot the comet—that honor was claimed by a French peasant who happened to be looking in the right place at the right time. Messier, however, was enthralled by the comet when he finally saw it, and resolved to make comet-hunting his life's work. Messier did, indeed, discover lots of comets, although that's not what secured him his place in the history of astronomy. Messier is famous not for the catalogue of what he was looking for, but for the catalogue of things that got in his way.

A distant comet appeared as a fuzzy object in the telescopes of Messier's era, as one would appear in a low-powered telescope of today. Comets are not the only fuzzy objects visible through a telescope, however, as Messier discovered. The chief difference between comets and the *other* fuzzy objects was that comets moved and the other fuzzy objects didn't. Messier began recording the locations of the other fuzzy objects so that he would not make the mistake of confusing them for comets. Messier's list of OFOs eventually grew to more than a hundred entries.

There was considerable speculation as to the nature of the OFOs. The brilliant French mathematician and physicist Pierre-Simon Laplace thought that they were distant clouds of gas; in fact, the word *nebula*, Latin for "cloud," was used to describe them. The philosopher Immanuel Kant propounded a competing theory, suggesting that nebulae might be vast congregations of stars so distant that the telescope was unable to resolve individual stars. Kant referred to these as "island universes."

The OFOs caught the attention of the English astronomers (and brother and sister) William and Caroline Herschel, who were at the cutting-edge of late eighteenth-century telescopy. Using data gathered from a giant telescope he had built, they published the *Catalogue of One Thousand New Nebulae and Clusters of Stars*.[1] This pair was the first to actually determine that one of Kant's "island universes" was indeed a vast congregation of stars, that collection being the Milky Way galaxy, the island universe in which our solar system is located. William Herschel actually mapped the Milky Way, and positioned the solar system in it near the center. Nevertheless, Herschel's telescope could not settle the question of what the nebulae were. The answer was to come as a result of the discovery of spectra by Bunsen and Kirchhoff.

In 1863, the British astronomer Sir William Huggins equipped his eight-inch telescope with a spectroscope and studied the spectra of stars other than the Sun. These, too, exhibited the same spectra as elements known on Earth. The following year, Huggins decided to look at the spectrum of a circular nebula in the constellation Draco. He was astonished to see that the spectrum consisted of a single bright line, which he recognized as corresponding to the spectrum of the element hydrogen. Nebulae were indeed, as Laplace had conjectured, distant clouds of gas! Shortly thereafter, Huggins observed the spectrum of the constellation Andromeda. That spectrum revealed the myriad lines seen when looking at a star. Nebulae were indeed, as Kant had hypothesized, gravitationally bound collections of stars!

Once astronomers knew what nebulae were, the next question was obvious: Where were they? Did they belong to the Milky Way galaxy, providing additional evidence that the Milky Way galaxy was the whole universe, or did they lie outside it? The answer was to have major ram-ifications for some of the biggest questions that had ever been asked.

The Period-Luminosity Relationship

The first measurement of the distances to the stars was done by using the idea of parallax. You can get an idea of how this works by looking at the minute hand of a distant clock at noon. If you look at the clock from left of the clock, the minute hand will appear to be between the 12 and the 1. Now move to the right of the clock, the minute hand will appear to be between the 11 and the 12. If you measure the distance you have moved to your right (the baseline distance) and the angles from which you are observing the clock, you can use trigonometry to compute the distance to the clock.

The problem with this method is that its utility is limited. Even using telescopes mounted on satellites orbiting our planet, which is now being done, the baseline distance is limited to the maximum separation of two points in the Earth's orbit (the distance between where the Earth is relative to the Sun on January 1 and where it is on July 1). At some stage the distance to a star is so great that the angles become too small to be measured; practically speaking, the parallax method only works for stars within a few hundred light-years of Earth. The fact that there were stars for which this method did not work implied that some of the stars were thousands of light-years away—and maybe even farther than that.

The first giant leap beyond the limits of parallax was taken by Hen-rietta Swan Leavitt, who had graduated from Radcliffe College in 1892. Nowadays, graduates of Radcliffe College, such as the most recent member of the Supreme Court, Elena Kagan, can look forward to many enticing employment opportunities, but there were very few openings for women in 1892. As a result, Leavitt took a job as a computer at the Harvard College Observatory, at a time when *computer* meant a person who computes. For the princely sum of $10.50 per week, she would measure the brightness of stars as they appeared in photographic plates.[2]

Although most stars maintain the same intrinsic brightness for millions—or even billions—of years, the so-called variable stars fluctuate significantly in brightness over brief periods. This was known as far back as 1638, when the astronomer Jean Holward[3] observed that the star Mira fluctuated in brightness in a cycle lasting eleven months. Although some stars fluctuate in brightness because their light is periodically dimmed by another celestial object passing between it and Earth, the Cepheid variables (first observed in the constellation Cepheus) pulsate due to a gas dynamics mechanism first elucidated by Sir Arthur Eddington.[4] The variable stars, especially the Cepheid variables, attracted Leavitt's attention. She noticed that the brighter the star, the longer its period, and in 1908 published a note to this effect in the *Annals of the Astronomical Observatory of Harvard College*.[5]

This note did not set the astronomical world afire, but Leavitt was not discouraged. She continued working with these stars, and in 1912 published what is now known as the period-luminosity relationship. Based on a study of 1,777 stars, she concluded, "A straight line can be readily drawn among each of the two series of points corresponding to maxima and minima, thus showing that there is a simple relation between the brightness of the variable and their periods."[6]

This epochal discovery, like Leavitt's earlier note, went largely unnoticed. Leavitt continued to work at the Harvard Observatory, and was promoted to the head of the photographic section by Harlow Shapley in 1921, only for her life to be cut short by cancer later that year. Solon Bailey, a colleague at the observatory, gave the following tribute to Leavitt at her funeral. "She had the happy faculty of appreciating all that was worthy and lovable in others, and was possessed of a nature so full of sunshine that, to her, all of life became beautiful and full of meaning."[7] It would not be for a little while longer before Leavitt's contributions to astronomy were given their due.

The period-luminosity relationship Leavitt had discovered enabled astronomers to gauge the distance to the Cepheid variables, providing a way to measure distances substantially in excess of the limitation imposed by the parallax method. The idea is easily illustrated using automobile headlights. Most automobile headlights are manufactured with a uniform brightness. Because we know how bright the lights are when a car is fairly close, we can tell how far away the car is by comparing how dim the light appears to its known brightness when the car is close by. Of course, we don't use simply this as an indicator as to when we can cross a street in front of an oncoming car; even at night in an area with no stoplights, there are other cues such as sound to guide us. Nonetheless, the principle is valid.

In the language of astronomy, the uniform brightness of manufactured automobile headlights provides a standard candle relative to which we can calibrate the brightness of other lights. The period-luminosity relationship discovered by Leavitt showed that the Cepheid variables could serve as standard candles as long as the brightness of one Cepheid variable could be determined (analogous to measuring the brightness of a single automobile headlight). It would not be necessary to perform the difficult task of measuring the brightness of any other, because its period (the timescale over which it went through a cycle of dimming and brightening) was easy to determine and its brightness could then be computed according to the period-luminosity relation.

Luminosity fades in a predictable way as the distance to the object decreases, so if it were possible to use parallax to measure the distance to a single Cepheid variable, then one would know the relationship between distance and luminosity for that star. Given any other Cepheid variable, one could use the period-luminosity relationship to deduce its luminosity; from its luminosity one could deduce its distance. Within a year of Leavitt's publication of the period-luminosity relationship, Ejnar Herztsprung—though probably not motivated by Leavitt's discovery but pursuing a research program of his own—had determined the distance to several Cepheid variables in the Milky Way.[8]

You have to give Shapley credit: he recognized the value of this technique, and made extensive use of the period-luminosity relationship to determine the size and shape of the Milky Way. However, you also have to debit Shapley for what I consider to be behavior more appropriate to a politician than to a scientist. In 1926, the Swedish mathematician Gosta Mittag-Leffler would contact Shapley about the possibility of nominating Leavitt for the Nobel Prize.[9] Mittag-Leffler was unaware that Leavitt had died, and so, because the Nobel Prize was (and is) awarded only to living scientists, she was ineligible. Shapley's next move was shameful: not to mourn hard work recognized too late, but instead to attempt to persuade Mittag-Leffler that the credit belonged not to Leavitt for the discovery of the period-luminosity relation, but to Shapley himself because of Shapley's use of the relation in determining the size of the Milky Way.

The Realm of the Nebulae

Still unsettled, though, was the location of the nebulae. There had been a serious attempt to settle it in 1920, when, in a classic debate at a meeting of the National Academy of Sciences, Shapley squared off against the astronomer Heber Curtis on the scale of the universe. The great contests pit not just opponents, but opposing styles, and Shapley was a rough-hewn plain-spoken type, whereas Curtis was an urbane patrician. Shapley took the side that held that the Milky Way galaxy constituted the entire universe, whereas Curtis held to the position that some of the nebulae lay outside the Milky Way. I wish that this debate had occurred during an era in which video recordings could have been made, because it must have been fascinating. Less so was the undercard, a presentation on the hookworm that history remembers only for having been dreary and long enough to bore Albert Einstein. So unimpressed was he that he remarked to another audience member, "I have just got a new theory of eternity."[10] The Great Debate, as it was later to be called, ended in a draw, because neither Shapley nor Curtis could prove their position. Shapley was faced with the difficult task of proving a negative, that no nebula lay outside the Milky Way, and Curtis, whose position would have been the easier to prove, simply lacked the data.

The final word on the subject would belong to the hero of this chapter, Edwin Hubble, a larger-than-life figure and one of the giants of twentieth-century astronomy. Tall, handsome, an accomplished athlete, a Rhodes Scholar, and a major in the U.S. Army during the First World War, Hubble joined the staff of the Mount Wilson Observatory in California in 1919. Later, Hubble married Grace Burke, an elegant woman and the daughter of a successful banker. There are several billion human beings, and probabilistic considerations dictate that some of them will have it all. Edwin Hubble was unquestionably such an individual.

Hubble is best known for two key achievements, the first of which was to resolve the Great Debate. During 1922 and 1923, Hubble discovered several Cepheid variables in what was then called the Andromeda nebula. Using Hertzsprung's calibration and the period-luminosity law, it was apparent that the distance to these stars was vastly greater than the size of the Milky Way, and so Andromeda had to be a collection of stars that lay well outside of the Milky Way. As a result, the Andromeda nebula is now known as the Andromeda galaxy, and the work of Hubble and those that followed showed that the universe is composed of billions of similar galaxies. Hubble also made the discovery with more grace than Shapley could have; Hubble had a deep appreciation of the importance of the period-luminosity relationship and Leavitt's role in discovering it, and felt that Leavitt should have won the Nobel Prize for it.

The Cepheid variable yardstick, like the parallax yardstick, has limitations, however, and Hubble and his colleagues would draw on another important discovery of nineteenth-century physics to show just how large the universe actually is.

The Doppler Effect

That discovery is the Doppler effect, which you may know from a policeman's radar gun or the weather report. The Doppler effect was first described by the Austrian physicist Christian Doppler,[11] who noticed that the tone of a train's whistle shifted as it went by. The effect was originally confirmed by one of the more picturesque experiments in the history of physics. Accurate measurement of the frequency of sound waves by instruments such as oscilloscopes did not exist in the 1840s; the most accurate instruments available at the time were the ears of trained musicians. To confirm the Doppler effect, musicians were installed on a train and told to play the same note. Other musicians, preferably possessing perfect pitch, were located on the side of the tracks and asked to determine the pitch of the sound they heard as the train passed by. Sure enough, the perceived note was higher than what the musicians were told to play as they approached, and lower as they went away.

The math involved is just simple algebra. Imagine you're standing in a railroad crossing with a train approaching at 25 meters per second, or about 55 miles per hour. To warn you, the engineer blows the whistle, which happens to be tuned to a middle C, which has a frequency of 260 hertz, or sound waves per second. Now, the important number here is the ratio of the speed of sound, which is 340 meters per second at sea level, and the frequency; that number for a middle C is 1.31 meters, and is the note's wavelength. Because the train is moving, however, you don't hear a middle C; the note you do hear is equal to the net speed of the sound divided by the wavelength of the note. So when the train approaches, you hear (340 meters per second + 25 meters per second) / 1.31 meters, or 279 hertz; when it is moving away, you hear (340 meters per second − 25 meters per second) / 1.31 meters, or 240 hertz. To be musical about it, you hear a D-flat coming and a B going away. We can relate the observed (Fo) and emitted (Fe) frequencies and the velocities of the train (V) and sound (v) like this, V being negative as the train approaches you:

$$Fo / Fe = 1 - V / v$$

The Doppler effect, as described above, works well as long as we are dealing with things like train whistles that travel at velocities vastly less than that of light. However, if you're dealing with a source of electromagnetic waves (which move at the speed of light), and the source of the waves itself is also moving at a reasonably large fraction of the speed of light, Einstein's special theory of relativity shows that there is a correction that must be made.

Time Out: The Special Theory of Relativity

One of the first (of many) indications that I wasn't cut out to be a physicist was that I had a great deal of difficulty understanding time dilation and length contraction, key features of the special theory of relativity. I've been a physics wannabe all my life, but I have never understood the great ideas of physics with the same clarity that I understand some of the great ideas of mathematics. Most of us have felt that we understand something incompletely at one time or another; comprehension flickers on and off as we think we've got it, and then we don't. Einstein's theory of special relativity has occupied such a position for me for the five decades since I was first exposed to it. I've read many of the great popularizations of it, as well as fully mathematical treatments in a variety of texts, and I never quite got it—until David McKay came up with the explanation I am about to present. It's the simplest and most straightforward of any of the explanations with which I am familiar.

If you have done a reasonable amount of train travel, you've probably experienced the following situation. There's a train on an adjacent track, and through the window you can see the other people on the train. All of a sudden, there is motion; you see the other train moving. However, unless there is something obvious to make you realize *which* of you is moving—such as a sudden jerk in your train or the motion (or lack thereof) of obviously fixed entities such as railroad ties or trees—you don't know whether you're moving or they're moving.

Relativity deals with two "frames of reference"—the two trains—that are moving at a constant velocity with respect to one another. Which one is *really* moving? Both! The key to the mathematics of special relativity is that if a person in either train measures the same distance in meters or the same amount of time in seconds, they will both come up with the same number. In order to understand the rationale for this, imagine that while the other train was next to yours, you had two meter sticks, and handed one to an individual in the adjacent train. If you measure the same distance in the outside world, you with your meter stick and he with his, you will get the same number because the question of who has the real meter stick (or—who is moving and who isn't) is moot; there is no way to tell because each sees the other moving.

So that's what you do: you open a window and hand a meter stick, a clock, and a flashlight to a person in the train just as that person (who also conveniently has an open window) is opposite you. It's a special flashlight; when you turn it on, it emits a single photon. The trains move away from each other at a constant velocity, which you measure as v meters per second; the person in the other train computes the velocity at which you are moving away from him as v meters per second as well. That's the basic assumption of relativity—neither of you has the "correct" meter stick and the "correct" clock while the other has the "incorrect" ones, and so each person must obtain the same numerical measurement, otherwise one measuring system is somehow "correct" and the other one isn't.

Now suppose that, at the moment the person in the train passes you, he aims the flashlight at the ceiling and turns on the flashlight; shortly thereafter the single photon hits the ceiling in his train. You construct a right triangle with three vertices. Vertex A is where you are at the moment the other person turned on the flashlight. Vertex C is the point on the ceiling of the train at the moment the single photon hits it; since the train has moved some distance down the track from the moment the flashlight was switched on, C is at the same level as the ceiling but further down the track. Let B be the point on the track directly below C; AC is the hypotenuse of a right triangle. We've known since Pythagoras that $AC^2 = AB^2 + BC^2$.

Distance equals rate × time, and c, the velocity of light, has the same numerical value (like the velocity of the train) whether you measure it or the person in the train measures it. How much time has elapsed between the moment the light is turned on and the moment the photon hits C? You *measure* distance, but *compute* time from the equation distance = rate × time. You denote this amount of time by T, and so $AC = cT$, since AC is the distance the photon has traveled; it started at A and ended up at C. Since the train is traveling at a constant velocity v, it started at A and ended at B, so $AB = vT$.

The person in the train denotes by t the amount of time that has elapsed between the moment the light is turned on and the moment the beam hits C. The photon travels from the floor of the train to its ceiling, the distance BC; the person in the train computes that distance as ct. So would you, because the line segment BC is perpendicular to the line segment AB, and both your meter sticks are unaffected by motion in the AB direction. We substitute these values—from your frame of reference—into the Pythagorean Theorem.

$$AC^2 = AB^2 + BC^2$$
$$(cT)^2 = (vT)^2 + (ct)^2$$

And so

$$t = T\sqrt{1 - \frac{v^2}{c^2}}$$

This is the famous "time dilation" effect; the number of seconds that have elapsed on the clock of the person in the train is less than the number of seconds that have elapsed on your clock. Notice that we didn't even need the clocks; we obtained the times from the distance = rate × time formula.

If we denote the distance AB as measured by you as L and the same distance as measured by the man in the train as l, we see that $L = vT$ and $l = vt$, because both parties see the velocity of the train as having the same numerical value v. So $v = L / T = l / t$, and therefore $l = (t / T)L$. Substituting the previously obtained value for t / T yields

$$l = L\sqrt{1 - \frac{v^2}{c^2}}$$

This is the equally famous FitzGerald contraction: the distance measured by the person in the train is less than the distance you measure.

The Special Theory of Relativity and the Doppler Effect

The special theory of relativity impacts the Doppler effect, because when things are moving close to the speed of light, the Lorentz factor (the square root quantity in the above equation) becomes significant; for small velocities it has a value that is basically indistinguishable from 1. The equation for the ratio of observed and emitted frequencies becomes this:

$$F_0 / F_e = \sqrt{\dfrac{1 - \dfrac{v}{c}}{1 + \dfrac{v}{c}}}$$

If $V = .9c$ (that is, the source is moving at 90 percent of the speed of light) and Fe equals 6.5×10^{14} hertz (that is, the emitted light is blue), then the observed light will have a frequency of either 1.49×10^{14} hertz (infrared) or 2.83×10^{15} hertz (far ultraviolet), depending on whether the light source is moving away from or toward the observer. Thus, when a light source moves away from us, its color is said to be red-shifted; when a light source moves toward us, its color is said to be blueshifted.

Doppler didn't predict many of the modern-day uses of the Doppler effect—not even using it to determine the speed of a pitcher's fastball—but he absolutely hit it out of the park when he said the following:

> It is almost to be accepted with certainty that this will in the not too distant future offer astronomers a welcome means to determine the movements and distances of such stars which, because of their unmeasurable distances from us and the consequent smallness of the parallactic angles, until this moment hardly presented the hope of such measurements and determinations.[12]

It is this use of the Doppler effect that has enabled us to answer the third of the deep questions that children frequently ask—where did it all come from?

The Expanding Universe

Using Cepheid variables as standard candles showed that there were many galaxies lying outside of the Milky Way. However, larger telescopes were being constructed, and the larger they were, the more objects they could see. Many nebulae were either devoid of Cepheids or may have had Cepheids that were undetectable, and so another method of determining distances needed to be found. The Doppler effect, with its relation of change in frequency to velocity, proved to be just the ticket (an appropriate choice of words, considering its use in documenting violations of speed laws).

When an object moves in a straight line, an observer can describe its velocity in terms of two numbers. Imagine that a car is moving at 60 miles an hour in the northeast direction along a line inclined at 60 degrees east of due north. In one hour the car will have moved 60 miles along the hypotenuse of a right triangle whose north-south side is 30 miles long and whose east-west side is $60^{\sqrt{3}}\!/_2 \approx 52$ miles. Physicists describe this process as the resolution of velocity into perpendicular components. When an object moves in space, the Doppler effect enables us to determine the velocity component along a line from the object to the observer; this velocity component is called the radial velocity.

Combining the Doppler effect with the spectroscope enabled astronomers to see how characteristic patterns of spectral lines in stars (or galaxies) changed; this change in the frequencies of the lines enabled the radial velocities of the stars (or galaxies) to be determined. William Huggins, the British astronomer who had solved the question of the composition of the nebulae, first did this for the star Sirius in 1872. He found a slight redshift in the lines associated with the element hydrogen, and careful measurement enabled him to conclude that Sirius was moving away from us with a radial velocity on the order of 47 kilometers per second.[13]

Over the next few decades, measurements were made on the radial velocities of a large number of stars; some were found to be approaching us, whereas others were receding. This did not surprise astronomers; after all, if you find yourself on a downtown street in the middle of the afternoon, some people will be walking toward you and others will be walking away from you. Unless some event has occurred, such as the sudden appearance of a major celebrity, to cause the majority of people to walk in a particular direction, people are just as likely to be walking away from you as toward you. Some will walk more slowly, others more rapidly.

Astronomers also tried to determine the radial velocity of nebulae. The spectral lines of many nebulae were so close together that they basically formed a continuous band, making it difficult to determine individual spectral lines. Without the ability to distinguish individual lines, the characteristic pattern of lines associated with particular elements could not be resolved, and so early determination of radial velocities was limited to stars. Typically, these velocities were in the range of about 10 kilometers per second.

By the twentieth century, technology had improved enough that the individual lines in the spectra of nebulae could be distinguished. This made it possible for Vesto Slipher, a young American astronomer, to determine the radial velocity of the Andromeda nebula. Even though this was prior to Hubble's use of Cepheids to determine the distance of the Andromeda nebula, Slipher's work resulted in an amazing result—the Andromeda nebula was approaching the Earth with a radial velocity of an astounding 300 kilometers per second; one-tenth of one percent of the speed of light. Slipher continued to work on the problem of determining the radial velocity of nebulae, and at the 1914 meeting of the American Astronomical Society he presented the radial velocities of a number of nebulae, which earned him what is now referred to as a standing O. Curiously, the vast majority of the radial velocities were recessional; the nebulae were apparently fleeing away from the Milky Way. Were these nebulae actually fleeing away from the Milky Way, or was there something out in deep space attracting them? No one had any idea.

One of the attendees at Slipher's 1914 presentation was Edwin Hubble. Hubble was able to use Leavitt's period-luminosity relation to determine the distance to a number of galaxies housing Cepheid variables, and could use Slipher's data, as well as his own, to determine the radial velocities of those galaxies. Here is the data that Hubble had available for his original analysis. A megaparsec is a unit of distance for measuring great distances, such as those between galaxies, and equals roughly 3.2 million light-years.

Hubble noticed, as you might also, that as the distance from Earth increased, so did the recession velocities. Plotting this data on a rectangular coordinate system, he noticed that the data seemed to fall approximately on a straight line. There is a standard technique from

DISTANCE FROM EARTH IN MEGAPARSECS	RECESSION VELOCITY KM/SEC
0.032	170
0.034	290
0.214	−130
0.263	−70
0.275	−185
0.275	−220
0.45	200
0.5	290
0.5	270
0.63	200
0.8	300
0.9	−30
0.9	650
0.9	150
0.9	500
1	920
1.1	450
1.1	500
1.4	500
1.7	960
2	500
2	850
2	800
2	1090

statistics used to determine what is called the regression line; it is the straight line that most closely fits a set of data points. In the above example, if the distance is given by x and the recession velocity by y, then the best-fitting line has the equation $y = -41 + 454x$.

Statistics also has a way of measuring how well data fit a straight line, called the correlation coefficient. If a data set perfectly fits a straight line with positive slope, the correlation coefficient is 1. If it perfectly fits a straight line with negative slope, the correlation coefficient is −1. If the data is essentially random, the correlation coefficient is 0. For the above data set, the correlation coefficient is 0.78; a pretty good fit to a straight line with positive slope.

However, Hubble also noticed that all the negative recession velocities were associated with galaxies that were relatively nearby. We now know the reason for that; galaxies cluster together in large gravitationally bound aggregations, and the nearby galaxies belong to a cluster known as the Local Group. The gravitational attraction accounts for the negative recessional velocities.

Hubble continued to augment his data set. As he began to obtain data for even more distant galaxies, he became convinced that there was a simple relationship between the recession velocity and the distance of the galaxy. This relationship, now known as Hubble's law, is expressed as $V = H_0 D$, where D is the distance to the galaxy in megaparsecs and V the recession velocity in kilometers per second. H_0 is the Hubble constant; although in the above equation it is given in (kilometers per second) per megaparsec, we could express D in terms of kilometers rather than megaparsecs and think of H_0 as being measured in units of 1 / seconds (or per second, or s^{-1}). The best estimate we currently have of H_0 comes from measurements made in 2010 with the Hubble Space Telescope, appropriately named after you-know-who. H_0 is approximately 71 (kilometers per second) per megaparsec, or 2.3×10^{-18} s^{-1}.

What could explain this amazing discovery? What model of the universe would account for the fact that the more distant the galaxy, the faster it is receding from Earth? Moreover, according to Einstein's general theory of relativity, there is no special place in the universe. Since the time of Copernicus, we have known that Earth is not the center of the universe, so not only must the galaxies be receding from Earth according to Hubble's law, they must be receding from one another.

Einstein's equations in the general theory of relativity suggested a possible solution—that the universe itself is expanding, and the galaxies are being carried along on the tide of expanding space. The more space there is between galaxies, the more expansion takes place, and galaxies that are further apart move away from each other more rapidly.

Hubble's law also enables us to obtain an answer to the question: how large is the visible universe? We use the term "visible universe" because if there is something out there with a recession velocity greater than that of the speed of light, we'll never know about it. Unless, of course, there's a way for it to emit information that travels faster than light—and if that's the case, there are going to be one hell of a lot of physical theories in need of change. We can certainly answer the question of how far away a galaxy would have to be in order to be receding from us at the speed of light. V would be 300,000 kilometers per second, and substituting 71 for H_0, we see that D would be 300,000 / 71 = about 4,225 megaparsecs, or 13.8 billion light-years.

The Size of the Visible Universe

In Chapter 9, we constructed a scale model of the universe using a recently purchased (and now eaten) grapefruit to represent the Sun. On that scale, 1 light-year is about 660 miles. The Milky Way galaxy looks somewhat like the pattern of water sprayed out by a rotating lawn sprinkler; it has a core at the center and several arms curving out from the core. The Solar System is located toward the end of one of those arms, and the Milky Way galaxy itself is about 100,000 light-years in diameter. Using our model, the distance from the Solar System diametrically across the Milky Way galaxy would be on the order of 60 million miles or so. If the Sun, the planet Mercury, and the Earth lie on a straight line with Mercury between the Sun and the Earth, the distance between the Earth and Mercury is approximately 60 million miles.

The Milky Way galaxy is part of a gravitationally bound collection of galaxies known as the Local Group. The Local Group is approximately 10 million light-years in diameter, and its center of mass is somewhere between the Milky Way galaxy and the huge Andromeda galaxy (formerly known as M31, or Messier 31, the 31st OFO on Messier's list), which is about 2,500,000 light-years from the Milky Way galaxy. We could therefore estimate the distance to the edge of the Local Group from the Milky Way galaxy as about 4,000,000 light-years. In our model, that would be a distance of 2,600,000,000 miles.

If the Sun, the Earth, and the planet Neptune were lined up with the Sun between the Earth and Neptune, and if Neptune were at its closest to the Sun, the distance between the Earth and Neptune would be just about right to represent the distance to the edge of the Local Group from the Milky Way.

The Local Group is itself part of a gravitationally bound collection of galaxy clusters known as the Virgo Supercluster. The Virgo Supercluster contains more than one hundred galaxy clusters, and the Local Group is about 60 million light-years from the furthest member of the Virgo Supercluster. Our model would therefore position the furthest galaxy cluster in the Virgo Supercluster as about 40,000,000,000 miles from Earth. It would take light about 2½ days to travel that distance.

The Virgo Supercluster is one of millions of superclusters throughout the visible universe. Our largest telescopes enable us to see close to 14 billion light-years out, which is close to the time of the big bang. Using our model, this would be about 1½ light-years from Earth. In the early 1990s, the domain of the Solar System was extended to include the Oort Cloud, a huge mass of icy bodies approximately 1 light-year from Earth. These are still gravitationally bound to the Sun, but our model goes 50 percent further than the Oort Cloud—approximately one-third of the distance to the nearest star.

The 14 billion light-year mark represents the "end" of the visible universe. Beyond this point, if Hubble's law continues to hold, the recession velocities exceed the speed of light, and so there is no way for information to get from whatever lies beyond this limit. There may be one hell of a party going on, but we'll never receive an invitation.

It is the universe itself that is expanding, carrying the distant galaxies like an ocean wave carries flotsam, and so it is no contradiction to have recession velocities greater than the speed of light. Recession velocities are not actual velocities, because the redshift of the galaxies comes not only from the Doppler effect, but also because space itself is expanding, and it is expanding faster further away from us. At any rate, I'm just hoping to live long enough to see the day that scientists determine whether there is anything beyond the limit imposed by Hubble's law, or whether what we can see is really all there is.

CHAPTER 13

OMEGA

I t's fitting that the final chapter of this book should deal with the constant that describes the ultimate fate of the universe. It's fitting because not only is omega the last letter of the Greek alphabet, but it's the constant whose value, once known, will tell us whether we live in a universe that is destined to expand forever or to collapse—and possibly be reborn. Of course, as a mathematician, I'm a huge fan of symmetry, so there's some pleasure having this book, which began with a chapter on gravitation, end with one as well.

Most people these days—and probably in any era that wasn't completely wretched—are present-day-centric, believing even if the current era is not yet the best of all possible worlds, it's superior to the other eras that have come before it. I certainly am. I like having the Internet, cable TV, Chinese takeout, painless dentistry, and air travel, although I long for the days when you could just show up and get on the plane. I'm also astounded by how much we know, and how much we have learned about the universe in the short span of time since my birth. Our knowledge is so vast, and our tools for finding and using it are so powerful, that it can be difficult to imagine how anyone could get anything done, or right, without them.

Einstein's General Theory of Relativity

One of the prevailing impressions of Einstein's theory of general relativity is that it is so incredibly difficult that at one time, only a dozen people understood it. In fact, when Eddington was asked what he thought of the statement that only three people understood the theory, Eddington is reputed to have wondered who the third person was.[1] George Ellery Hale, the director of the Mount Wilson Observatory, helped advance this view when he said, "The complications of the theory of relativity are altogether too much for my comprehension."[2]

That was then, but this is now, and Steven Weinberg, winner of the Nobel Prize for Physics and author of *The First Three Minutes*, disagrees. He writes, "It never was true that only a dozen people could understand Einstein's papers on general relativity, but if it had been true, it would have been a failure of Einstein's, not a mark of his brilliance."[3] General relativity is now a standard subject in the physics curriculum, which naturally suggests that even though it is difficult to master, most of the physicists of Einstein's day were capable of mastering it too.

Nevertheless, the Einstein field equations (EFE), which are the mathematical heart of the general theory of relativity, do present a real mouthful to the uninitiated. Take, for example, Mathworld's succinct description of them: a system of sixteen coupled, nonlinear, hyperbolic-elliptic partial differential equations.[4] Okay, that's a mouthful and a half—but it is a digestible mouthful and a half to an upper-division undergraduate in any one of several majors. Some of the terms are easy to understand, others require calculus—but even without knowing calculus it is possible to get a feel for what the EFE are.

First of all, they are equations, but they are not like ordinary equations most people are familiar with, such as $2x + 5 = 7$, the solution to which is a number. They are equations describing the rates at which certain parameters change at different places and different times— that's the "partial differential equations" part—and their solutions are functions, rather than numbers. The "hyperbolic-elliptic" part simply describes a particular type of partial differential equation, much as "quadratic" describes a certain type of single-variable equation.

"Coupled" merely means the variables often appear together in the same equation. For example, in the two equations on the next page, the variables x and y are not coupled.

$$2x + 5 = 7$$
$$3y - 1 = 8$$

However, in the following equations, the variables x and y are coupled.

$$2x + y = 5$$
$$7x - 2y = 1$$

Anyone familiar with high-school algebra knows that the two un-coupled equations above can be solved separately, but that to solve the coupled set requires working on them together, using an algebraic technique such as elimination. Coupled equations are almost always more difficult to solve than ones that aren't.

Finally, a linear equation is one such as $2x + 5 = 7$ or $2x + y = 5$, in which all the variables appear by themselves; they aren't raised to any powers, nor are functions such as logarithms applied to any of the variables. A nonlinear equation such as $x^3 + 5x = 18$ is always more difficult to solve than a linear one.

Upper-division undergraduates back in Einstein's day would have had no trouble decoding the idea of coupled, nonlinear, hyperbolic-elliptic partial differential equations. What made Einstein's work so significant was the generality of the phenomena that it described, and the depth of understanding necessary to derive the equations.

However, I don't think you need to understand any of this to appreciate the elegant simplicity of the "language" in which Einstein phrased his results. The EFE can be written simply as

$$G_{\mu\nu} = \frac{8\pi G}{c^4} T_{\mu\nu}$$

The single G on the right side of the equation and the c are the gravitational constant and the speed of light. The other symbols are tensors, which are simply condensed ways of writing a lot of information; the μ and ν have values from 0 to 3, where 0 represents the time coordinate ct (there's a technical reason that time t is multiplied by the speed of light) and 1 through 3 represent the three space coordinates. All four coordinates together describe a unique location in space-time, a specific place (done with coordinates 1 through 3) and a specific time (done with coordinate 0). There are actually sixteen separate equations, corresponding to the sixteen different ways $00, 01, \ldots, 23, 33$ of choosing one value for μ from 0 to 3 and one value for ν from 0 to 3. However, there's some redundancy in these equations, and they can be reduced to just six.

The Possible Universes

Almost all math instructors have had the following experience with students encountering story problems—the student tells the instructor that he has no difficulty solving the equations, his problem comes in setting them up. The Einstein field equations are a totally different situation; the "story problem" of the universe is set up within the equations, the problem is to determine how that story unfolds by solving the equations. Several people tackled this problem when the general theory of relativity was published, and there were some missed tackles in the process.

The first person to take a shot at solving the equations was Einstein himself, as could be expected since he was the first person to know what the equations actually were. The general theory was published in 1916, before Hubble had even demonstrated that galaxies outside the Milky Way existed, to say nothing of the fact that they were receding. The prevailing viewpoint was that the universe as a whole was static and unchanging. Consequently, Einstein wanted to find a solution that was static and unchanging. He found a solution, but it wasn't a static and unchanging solution; the universe it described either expanded or collapsed.

What to do? Einstein was sure his equations were correct, as they gave the right results when applied to the solar system, but something must have gone wrong on a larger scale. So Einstein inserted a "fudge factor" into the EFE in order to obtain a solution that would give a static and unchanging universe. This fudge factor morphed the EFE from the beautifully simple

$$G_{\mu\nu} = \frac{8\pi G}{c^4} T_{\mu\nu}$$

to the not quite as simple, but still elegant

$$G_{\mu\nu} + \Lambda g_{\mu\nu} = \frac{8\pi G}{c^4} T_{\mu\nu}$$

The extra term consists of the cosmological constant Λ multiplied by a tensor that facilitates the computation of the length and angles between tangent vectors; it enables one to compute distances and visualize how the space the tensor describes is shaped. The effect of the cosmological constant is that it counteracts the disturbing (to Einstein) tendency of the universe in Einstein's original solution to either expand or collapse. Theory in physics (or anywhere else) is pragmatic; you try to construct a theory that fits the facts, and the universe as Einstein knew it in 1916 was static and unchanging.

Einstein's solution is not the only solution to the equations. Another was quickly discovered by the Dutch astronomer Willem de Sitter.[5] Unfortunately, like Einstein's original solution, de Sitter's seemed to suffer from some nonphysical characteristics.

For starters, de Sitter's universe, by assumption, contained no mass whatsoever. This isn't as great a defect as it may appear, because, as we have seen throughout this book, the actual universe that we inhabit contains just a measly atom per five cubic meters of space. Our local neighborhood might seem cluttered with mass, but if we consider a sphere, a little more than a light-year in radius, centered on our home planet, that sphere would have an average density of just one atom per cubic millimeter—pretty close to a vacuum, and we're talking about a region chock-full of matter, relatively speaking. De Sitter's assumption of an empty universe might have struck his rivals as nonphysical, but in the end it is a pretty good description of the way things are. Nevertheless, it wasn't especially welcome. The second apparent problem was that, in de Sitter's universe, clocks at great distances from Earth ran slower than those on Earth. This had the effect of redshifting light from the distant galaxies. The de Sitter solution was described before Hubble's law was propounded, and both de Sitter and Eddington worried that the redshifted light would be erroneously interpreted as recession velocities. Not surprisingly, Einstein was not thrilled by the deficiencies of the de Sitter solution, saying that it didn't make sense to him (remember that Einstein did not know at the time that there were galaxies other than the Milky Way). De Sitter pointed to the evidence that Slipher was uncovering with regard to recession velocities of stars, and emphasized that these supported his model. (To this I raise my own objection: how can recession velocities of material objects support de Sitter's solution, when there are no material objects in the de Sitter universe? I can't imagine anyone doing something like this at a math conference. My guess is that stuff like this makes cosmology conferences a lot more entertaining than math conferences.)

The Greatest Debate

Astronomy has had more than its share of great pitched battles, starting with Copernicus versus the Catholic Church on the nature of the solar system. If the face-off between Harlow Shapley and Heber Curtis as to the question of whether the Milky Way constituted the entire universe was known as the Great Debate, then the argument as to the origin of the universe that took place during the middle of the twentieth century should certainly be accorded the title of the Greatest Debate.

There are really only two possibilities for the origin of the universe. Either it had to start sometime (the idea of the big bang), or it's always been there (continuous creation). However, one of the major characteristics of science is that theories must be constructed to fit the facts, and Edwin Hubble had conclusively shown that the majority of galaxies were receding from Earth, and were doing so in accordance with Hubble's $V = H_0 D$ law. In fairness to Hubble, he wrote the law as $V = KD$; the K was later changed to H_0 to honor his role in discovering it. In fact, it is generally considered to be bad taste in mathematics and science to name anything for yourself; the scientific community will do this if it deems your discovery sufficiently important. We speak of Einstein's theory of general relativity, but Einstein was not so brash as to call it that. The only case that I know of where a mathematician or scientist tried to jump the gun on such attribution was with Leonhard Euler, who spent a great deal of time developing ideas related to e, the base of the natural logarithms. Some historians feel that the letter e was so chosen because it is the first letter of Euler's last name.

So let's take a close look at the two contenders for the Theory That Explains the Origin of the Universe.

The Big Bang

In the early 1920s, there were three solutions to the EFE on the table. The first, Einstein's original solution without the cosmological constant, either expanded or collapsed. The second was Einstein's solution when an appropriately chosen cosmological constant was inserted, a static universe intended to fit the presumably known facts. The third was the de Sitter solution, which suffered from containing no matter and having redshifts produced by the slowing of time at a great distance from Earth.

The middle ground was a solution found by Georges Lemaître, a remarkable individual with a truly eclectic background. In addition to being an extremely talented physicist, Lemaître was also an artillery officer during the First World War who was decorated for bravery—and who later became an ordained Roman Catholic priest. As I've mentioned, systems of partial differential equations are remarkably difficult to solve, so it is not surprising that Einstein and de Sitter had not exhausted all the possible solutions of the EFE. Lemaître managed to find a solution of the EFE (with a cosmological constant included) that did more than simply describe an expanding universe—it supplied a mathematical derivation of Hubble's law. He published this paper, with the lengthy title of (translated from the French) "A Homogeneous Universe of Constant Mass and Growing Radius Accounting for the Radial Velocity of Extragalactic Nebulae."[6] This paper also provided the first computed value of the Hubble constant, derived from observed measurements. However, in physics as elsewhere, it's hard to get noticed when you're far from the center of the action, and this paper was published in an obscure Belgian journal, mostly because at the time Lemaître was an obscure Belgian professor of astronomy.

Amazingly, Aleksandr Friedmann, a Russian mathematician who had also served in the First World War, had independently developed the same approach, and had done so five years earlier than Lemaître. However, a lot of physics got done during the 1920s, and even though Friedmann published his paper in the prestigious German journal *Zeitschrift fur Physik*, he made what in retrospect turned out to be a poor public relations move; he titled his paper simply "On the Possibility of a World with Constant Negative Curvature of Space."[7] Friedmann's solution actually allowed for all three possible types of universes, with positive, negative, and zero curvature (as on the surface of a sphere, a saddle, and a sheet of paper, respectively). Maybe Einstein didn't see Friedmann's paper, or maybe the title persuaded him that this was not a paper worthy of his time. At any rate, Einstein didn't read it. It seems obvious in retrospect that he should have titled the paper "A Solution to Einstein's Field Equations." That would probably have got Einstein's attention!

Friedmann never did get Einstein's attention—he died of typhoid contracted on a visit to the Crimea in 1925—but Lemaître eventually did, in 1927. This was before the bulk of Hubble's work had been accepted, and Einstein, still holding to his vision of a static universe, was not immediately impressed. He said that while Lemaître's mathematics was correct, his physics was abominable.[8] However discouraging Einstein's reaction must have been, Lemaître continued to develop his theory. Hubble's data obviously provided a tremendous boost for his ideas, and Lemaître gained a strong ally in Eddington, who brought Lemaître's theory front and center by publishing a long commentary on it in 1930 in the *Monthly Notices of the Royal Astronomical Society*. Here at last was acceptance—Eddington described Lemaître's theory as a "brilliant solution" to the problem of the physical description of the universe—and it led to an invitation for Lemaître to present his theory in London: the big stage at last! It was on this occasion that Lemaître presented the idea of the universe expanding from an initial point, which he called the "primeval atom." Einstein, too, was ultimately won over by Lemaître's theory, and when, on a subsequent trip to California, Lemaître presented his work, Einstein stood up, applauded, and is reputed to have said, "This is the most beautiful and satisfactory explanation of creation to which I have ever listened."[9]

The Catholic Church loved it as well. How could it not? Not only was the author a priest, but as physicists explored the consequences of Lemaître's theory, they realized that if one wound back the clock far enough, the receding galaxies of today must at one time have emanated from an infinitesimally small volume, and that the universe began when this matter and energy was released. For a short period of time, the temperature of the universe was so high that it was impossible for atoms to form; the universe itself was pure energy. The Vatican was eager to attribute this to God, as in Genesis 1:3: "And God said, Let there be light, and there was light."[10] It was all part of the Vatican's long evolution from burning Giordano Bruno at the stake in 1600 to sponsoring conferences on astrobiology in the twenty-first century.

Do you find the idea of cramming the entire known universe into a volume substantially smaller than a pinpoint difficult to believe? I do, too. The Schwarzschild radius of the Sun is about 3 kilometers, and a typical galaxy contains billions, sometimes even trillions, of solar masses. The universe contains billions, possibly trillions, of galaxies. The big bang theory requires all that matter to be stuffed into a volume inconceivably smaller than a hydrogen atom, which itself is almost inconceivably small. As a mathematician, I can consider infinite-dimensional spaces with aplomb, because I know that they are merely mathematical constructs. But stuffing the entire universe inside this incredibly small volume? I guess I'm capable of Orwellian double-think on this issue: I can accept it because it is currently the best theory, according to experts, of how it all began. If I had read a mathematical development of this theory—which I shamefacedly confess I haven't—I'm sure I'd accept it even more readily. But as someone who has to sit on his suitcase in order to be able to close it before taking a trip, I find it impossible to believe from a strictly human standpoint. How can you get all that stuff in such a small place? It's not the big bang and what happened thereafter that bothers me, I just simply cannot conceive of that amount of mass in that infinitesimally tiny volume. Maybe someone will come up with a theory in which the contents of the current universe are stuffed inside a volume the size of the solar system—still hard to believe, but a lot more believable (at least for me), and the big bang starts the ball rolling from there.

I guess the distinguished astronomer Fred Hoyle, the man who actually coined the phrase "the Big Bang," had difficulty with this idea too—because he was even willing to sacrifice one of physics' most cherished principles in order to propound a theory that he felt was more credible.

Continuous Creation

Maybe Hoyle, and Thomas Gold, and Herman Biondi, did as poor a job of packing for a trip as I do, because they were bothered by certain aspects of the big bang theory. If you are going to come up with a competitive theory, however, you absolutely have to accept anything that has been confirmed empirically—and that meant that Hubble's redshifts needed to be explained. One school of thought, originally propounded by Fritz Zwicky, was that the redshifts really did not represent recessional velocities. Zwicky thought that one possible explanation might be "tired light"; light somehow lost energy on its journey from the galaxies to Earth, possibly as a result of its interaction with matter.[11] Another alternative theory was developed by A. E. Milne, who suggested that the universe was filled with galaxies with randomly distributed velocities (and moving in randomly distributed directions); the speed merchants among these galaxies would more likely be at considerable distance from us.[12]

Both these alternatives fell by the wayside, and Hoyle, Gold, and Bondi had to come up with a theory that allowed for Hubble's law and the expansion of the universe. In order to create a universe that looked the same at all times, thus eliminating the need for a big bang, they had to make sure that the universe was continually repopulated with stars and galaxies to replace those that had receded to the point where we no longer could observe them. That meant the principle of conservation of matter had to be scrapped, and that new matter had to be created. However, as the theory developed, the rate at which new matter needed to be created was unbelievably small, because the universe is so incredibly empty. Continuous creation was pretty close to nonexistent creation, as all that was required was the appearance of one new atom of hydrogen every billion years in each cubic meter of space.[13]

How could you ever verify—or disprove—such a theory? Obviously, we can't sit around observing a totally empty cubic meter for a billion years waiting to see that glorious instant when an atom of hydrogen appears. Additionally, there was the problem that not only did this method require the creation of hydrogen, it was also necessary to create a sprinkling of deuterium, helium, and lithium—because nucleosynthesis in stars could not produce these elements in the quantities in which they were observed. The theory was attractive, because it postulated a Universe beloved of symmetry fans, one that was isotropic and homogeneous in mass, space, and time—which is to say, it looks the same, no matter what direction you are looking or where you are. The theory was ugly, because it not only required one to dispense with the bedrock conservation principles that had never been observed to be violated, it required creation not only in the right quantities but in the right proportions.

And the Winner Is . . .

Not a big surprise—you already know that the winner was the big bang theory. It emerged victorious for several reasons. First, the big bang theory predicted a fundamentally universal temperature for the universe as a leftover cooling from the big bang itself; this was discovered in the early 1960s. Second, the big bang theory correctly predicted the abundances of various types of matter from fundamental assumptions that continuous creation required from ad hoc ones. Finally, the continuous creation theory required that the universe look the same at all times, and by the late 1960s, the discovery of quasars and pulsars at great distances was evidence that the universe was not the same at all times. Steven Weinberg effectively buried the continuous creation theory in a speech in 1972, but did so with a flattering eulogy. "Alone among all cosmologies," said Weinberg, "the steady state model makes such definite predictions that it can be disproved even with the limited observational evidence at our disposal."[14]

Friedmann's Solution, the Critical Density, and Omega

Friedmann's solution to the EFE may have escaped Einstein's attention when it was published, but it is at the core of the discussion of the universe's destiny. Deriving the solution from the EFE is far beyond the scope of this book—and I'm not altogether sure that the training that I have in mathematics would enable me to follow it—but the important thing is the result. If we assume that Friedmann's solution has been obtained—in which the universe could have positive, negative, or zero curvature—there really isn't a whole lot of work to find the value for the critical density that determines the ultimate fate of the universe. In fact, the math after that is so easy that some first-year algebra teacher could use it as an exam problem.

Friedmann started with the EFE as Einstein wrote it, including the cosmological constant.

$$G_{\mu\nu} + \Lambda g_{\mu\nu} = \frac{8\pi G}{c^4} T_{\mu\nu}$$

The fact that μ and ν take the values 0, 1, 2, and 3 may be likened to a table of equations with 4 rows and 4 columns. Freedman worked with the equation in Einstein's table of equations that had both μ and ν equal to 0. After what I suspect was a lot of work, he simplified this equation to read

$$H^2 = \frac{8\pi G}{3}\rho - \frac{kc^2}{a^2} + \Lambda\frac{c^2}{3}$$

H, G, and c are old friends: the Hubble constant, gravitational constant, and the speed of light, respectively. Λ is the cosmological constant, ρ is the average density throughout the Universe, a is a function of time called the scale factor, but can mercifully be taken to be equal to 1 today. Friedmann further simplified matters by assuming that $k = 0$ (which it turns out is what it seems to be, empirically), and that $\Lambda = 0$ as well. The above equation reduces to

$$H^2 = \frac{8\pi G}{3}\rho$$

This can be easily solved for ρ, obtaining the value of $\rho = 3H^2 / 8\pi G$ for the critical density; if we plug in the latest values for H and G, we get a critical density of about 5 atoms of hydrogen per cubic meter. This is just stunningly small—anyone's gut would say that the gravity created would be so small as to be basically nonexistent—but nevertheless it's big enough. This density is the critical density—the density of a universe that is flat and doesn't expand—and is usually abbreviated ρ_c.

Omega is simply the fraction whose numerator is the actual density and whose denominator is the critical density: $\Omega = \rho / \rho_c$. So, for Friedmann, the value of Ω would tell us whether the universe will collapse, which occurs if $\Omega > 1$, or expand. The trick is finding the density of matter in the universe.

Dark Matter, Radiation, and the Cosmological Constant

We're probably in a good position to determine exactly how much visible matter there is in the universe. Our telescopes are extremely good, we can pick up galaxies almost thirteen billion light-years away, and we can extrapolate densities to get a pretty good idea of the total amount of visible matter. As we noted earlier, there's only about one atom of hydrogen in every 5 cubic meters, only about 4 percent of the mass needed to reach critical density.

However, the bulk of the universe appears to be made up of dark matter. Ever since the 1970s, evidence has been steadily accumulating that every galaxy is surrounded by a halo of dark matter.[15] This evidence is in the form of the velocities of the distant stars in the galaxy around the center of mass of the galaxy; these velocities don't match up with what they should be if the only mass in the galaxy is what we see. What we see isn't what we get, gravitationally speaking. There's more mass in the galaxies than can be accounted for by visible matter.

The nature of the dark matter is the subject of considerable speculation. These theories range from the mundane (stuff that's familiar but isn't radiating, like really dark rocks), to the far out (supersymmetric particle theories that posit an entire new class of matter that no one has yet seen). However, that's for the physicists to worry about, the cosmologists (and those of us who breathlessly await the determination of the eventual fate of the universe) are just interested in how much of it there is. Is there enough of it to push Ω beyond 1, compelling the universe to collapse, or not?

The best estimate at the moment is that there isn't enough matter, light, dark, or anything else, to do the job. However, the matter density of the universe isn't all that matters—energy matters too. There's a lot of radiation in the universe; radiation is energy—and Einstein's special theory of relativity gives us the relation between matter and energy as $E = mc^2$, or $m = E / c^2$. So energy, too, enters the omega bookkeeping.

Finally, the cosmological constant is not actually zero—according to the latest measurements it is .7—which also acts to increase the density of matter and energy in the universe. The result of this is that the ratio Ω is somewhere between .98 and 1.1. The key number 1 lies within that range, and there's a fascinating argument that the universe is so fine-tuned that Ω is *exactly* equal to 1.

A Compelling Argument That $\Omega = 1$

I first came across the argument that Ω needed to be exactly 1 in Martin Rees's fascinating book *Just Six Numbers*, published in 2000. At that time, given the rates at which the Universe was known to be expanding, it had been shown that if, just one second after the big bang, Ω differed from 1 by as much as 10^{-15}, then Ω would differ significantly from 1 now. Think of how astonishing that is; Ω must have been somewhere between .999999999999999 and 1.000000000000001 one second after the big bang in order for it to be in the range it is now. Just six years later, in Frank Levin's *Calibrating the Cosmos*, he quotes the simply mind-blowing improvement that Ω must have been between $1 - 10^{-52}$ and $1 + 10^{-52}$ in order for Ω to be where it is now; I'm not going to further alienate my copy editor by writing down a decimal point with 52 nines following it for the lower limit.[16]

This is fine-tuning on a grand scale; how can one *not* conclude that $\Omega = 1$? However, our theories are simply incapable of measurements discerning this—but our theories *might* be capable of describing a mechanism that would conclude that Ω must be equal to 1. So far, no luck—but that's not so surprising. At present, we have only been able to conclude that the ratio of the magnitudes of the positive charge on a proton to the negative charge on an electron lies between $1 - 10^{-21}$ and $1 + 10^{-21}$.[17] Achieving a level of accuracy such as that is an astonishing bit of technical wizardry, and it would certainly surprise every physicist on the planet if that ratio were not equal to 1—but no one has come up with a theory that explains why this ratio should be exactly equal to 1. The best we can do so far is to conclude that almost certainly, it just *is*. Both Newton's law of universal gravitation and Coulomb's Law, the basic inverse-square laws on which much of physics rests, are fine-tuned; why should the exponent be 2 and not some number really close to 2 but not exactly equal to it? Of course, these laws appear to be tied in with the Euclidean geometry that underlies our universe, but mathematicians have devised a lot of non-Euclidean geometries—why should our universe have selected this precise one?

Is Omega the Whole Story?

Every so often, science discovers something that causes me to say, "Are you kidding?" This happened in 1998, when cosmologists announced that, based on a study of twenty-one Type-IA supernovas,[18] the very fabric of space was expanding. When I read this, I blinked. Twenty-one lousy supernovas and they were ready to totally revamp cosmology? In statistics, sample sizes of less than thirty are considered unreliable. What was going on?

This conclusion, which seemed like a giant leap to me, was based on an updated version of Henrietta Swan Leavitt's discovery of the period-luminosity relationship for Cepheid variables, which enabled Cepheid variables to be used as standard candles. The 1998 study used Type IA supernovas as their standard candles. In theory, Type IA supernovas explode in the same fashion, so their intrinsic brightness is known. Based on a study of twenty-one supernovas that appeared dimmer than they should have, cosmologists concluded that this was evidence for a positive cosmological constant and a hitherto-unsuspected dark energy (dark because we had no idea what it was) that was accelerating the expansion of space.

That expansion, if it indeed exists, is a game-changer. If some unknown energy is pushing the galaxies apart, as cosmologists currently think, it could overwhelm the gravitational force and so force the universe to expand even if the value of omega were such that gravity on its own would compel the universe to contract. Such an accelerating expansion of space goes by the catchy name of the "big rip," and catchy titles sell theories as well as books.

While this is currently the hottest theory going, it's not being written into the textbooks with the same level of confidence as the theory of relativity. This particular applecart could be upset; for one thing, Type IA supernovas might not turn out to be as reliable as standard candles as is currently thought. I keep thinking of Leavitt basing her conclusions on the study of 1,777 Cepheid variables; I bet if we could ask her what she thought, she'd be a lot more cautious, if only for statistical reasons.

It's unfortunate that Las Vegas is not especially interested in posting lines on the validity of scientific theories, because I could have made money betting against three headline-making developments of the past. I admit that I have no documented evidence to back my next assertion, but I didn't believe that water molecules had a memory,[19] that cold fusion could be achieved on a tabletop,[20] or that there was a fifth force operating exclusively at middle distances.[21] I'd like to see the line on the big rip, because even though it is the current fair-haired boy of cosmological theories, it just somehow feels wrong to me. But there is a theory that feels right.

My Favorite Universe

I've spent my life doing mathematics but reading about the sciences, and I know that I'll never be able to really appreciate some of the scientific theories I've read. I don't think you can truly appreciate a profound theory unless you really understand it, and I simply don't have either the background or the intellectual tenacity needed to comprehend some of the current theories of cosmology. Richard Feynman once remarked that if there was a deep theory that explained everything, that would be terrific, but if reality was like an onion and you simply peeled off one layer to expose a deeper layer, that would be terrific, too.

But not, in my opinion, as terrific. What I would like to see is a theory that reconciled the dynamic aspects of the big bang theory with the eternal nature of continuous creation, and fortunately there is a theory out there that does it.[22] It views our universe as a fragment of a multiverse, which is akin to a perpetually boiling pot of soup; continually forming new bubbles (each of which is a universe started by a big bang). The bubbles eventually grow and burst, but the pot of soup is eternal.

I watched a lot more television when I was younger. Two of my favorite shows were *M.A.S.H.* and *Magnum P.I.* Both of these shows had highly publicized final episodes—in fact, I think the final episode of *M.A.S.H.* is the most-watched TV show in history[23]—and to this date, I've never seen either of these. Ridiculous though it sounds, I felt I could watch unseen episodes and reruns of these shows as long as I hadn't seen the final episode—for that final episode would bring the curtain down and the characters would cease to exist.

I also read a lot of mystery stories, but when *Curtain,*[24] the Hercule Poirot mystery that Agatha Christie made sure was published posthumously, finally did come out, I never read it for the same reason. I mentioned this to my wife, and for my last birthday she gave me a copy of *Curtain.* It's sitting on my desk, and the next time I hop a plane, I'm going to start reading it. And maybe I'll watch the final episodes of *M.A.S.H.* and *Magnum P.I.* if they show up on the tube.

Some things simply have to come to an end. *M.A.S.H.* This book. Its author. Life. I can accept all of those—but not the end of the universe, because if the universe does not end, all the good things might continue. Or come again.

CODATA NOTE

The four-year cycles that involve the years 2002, 2006, 2010, etc. feature some highly publicized events. The Winter Olympics. The World Cup. Off-year elections. Garnering considerably less attention, but certainly of more long-lasting significance than sporting events—and perhaps even than off-year elections—is the quadrennial report of the Committee on Data for Science and Technology, known as CODATA.[1]

If the efforts of this group are ever made into a musical, it might be a good idea to see if the owners of the Rodgers & Hammerstein copyrights would allow the lyrics and tune for the opening number to *Annie Get Your Gun*[2] to be adapted for the purpose. Every time I read CODATA, the refrain "There's no data like CODATA, like no data I know" echoes through my head. For there really is no data like CODATA; it's the data on which the entire science community relies for its theories and experiments, and is the object of a continuing—and hopefully never-ending—search for improvements to the values of the fundamental (and not quite as fundamental) constants of the physical sciences.

The full 2010 report has not been published as of this writing; the most recent one that I have seen is the 2006 paper, a 105-page PDF available for downloading through the good offices of arXiv.org.[3] The *x* in *arXiv* is the Greek letter chi, so *arXiv* is pronounced the same as the word *archive*, and it is a huge resource of publications and prepublications for the scientific community. Alternatively, if you only want to look at a few of the constants, you can see them courtesy of the National Institute of Science and Technology.[4]

Obviously, you won't want to read the full CODATA set, but one notation may cause you a little confusion. The CODATA value for G is given as $6.67428(67) \times 10^{-11}$; for the purposes of this discussion, we needn't worry about what the units are. The two digits in parentheses go by the name *combined standard uncertainty*; it's basically an acknowledgment that when you take a lot of measurements, there's going to be a certain amount of scatter. 6.67428×10^{-11} is an average of these measurements, and the 67 that's in parenthesis is a statistically adjusted standard deviation whose value is 0.00067×10^{-11}.

Almost a century ago, before the discovery of X-rays opened the door to the subatomic world, some physicists believed that the job of physics was done, and all that remained to do was to tack on another digit or so to the physical constants. Looking at the CODATA report, it is clear that the business of tacking on more digits is a flourishing industry. I can think of at least three reasons this is worth doing, and I am sure there are more.

First, the practical aspect. Maybe you've never heard of 99942 Apophis; it's a rather hefty asteroid that at one time was judged to have almost the same probability of hitting the Earth as a gambler does of rolling snake eyes; this is crapping out, big-time. That probability is now well below one in a million, thanks to more observations—and better knowledge of the value of G. A more accurate knowledge of G will allow us to better assess the threat of an asteroid or comet hitting Earth.

Second, basic science is often the result of discovering that existing theories are wrong. A theory may be right to a number of decimal places, but a discrepancy might show up in the next decimal place. Better scientific theories mean better technology; better technology means a better life.

Third, the search for improvements in measurement often requires the creation of better technology to do the job, and if you read the previous paragraph, you know where I stand with regard to better technology.

Every so often, I'll see a story in the paper (which I still read assiduously every morning) that some computer somewhere has computed pi to a record number of decimal places, or that a new largest prime has been found. Largest, of course, in the sense that this is the largest prime we've found—it's been known since Euclid that there is no largest prime. I think these stories appear because setting records has a certain appeal. However, I can't ever recall seeing a story that scientists have just been able to determine the Planck constant—or any other—to yet another decimal place. I'm willing to grant that these events are probably not newsworthy, but for the three reasons I mentioned above—and for the others that I'm sure exist, improvements in measurement can result in measurable improvements in our lives.

NOTES

CHAPTER 1

1. D. Whiteside, "Sources and Strengths of Newton's Early Mathematical Thought" in *The Annus Mirabilis of Sir Isaac Newton, 1666–1966*, ed. R. Palter (Cambridge, MA: MIT Press, 1970), 74.

2. Ibid.

3. Ibid.

4. J. Gribbin, *The Scientists: A History of Science Told Through the Lives of Its Greatest Inventors* (New York: Random House, 2003), 181.

5. I. Newton, *Philosophiae Naturalis Principia Mathematica*, trans. Motte revised by Cajori (Berkeley: University of California Press, 1962).

6. T. Koupelis, *In Quest of the Universe* (Sudbury, MA: Jones & Bartlett Publishers, 2011), 62.

7. S. Hawking, *Principia, Isaac Newton: On the Shoulders of Giants* (Philadelphia: Running Press, 2002), 32.

8. The noted Italian astronomer Giovanni Cassini, whose namesake spaceship is currently orbiting Saturn and its moons, was the first to come up with an accurate measurement of the distance from the Earth to the Sun. He used what is known as the parallax method, which took advantage of the improved telescopes available in his era and a simple fact: if you observe a nearby object against a fixed background from two different positions, the object shifts against the fixed background (you can see this by looking at a nearby object against a distant skyline with just your right eye and then with just your left eye). Measurements of the angles involved and the distance between the observing positions, coupled with geometry and trigonometry, enable one to compute the distance to the nearby object. Cassini and a fellow astronomer took simultaneous measurements from Paris and from French Guiana, and Cassini arrived at a distance from the Earth to the Sun that differed by only about 1 percent from the accepted value today.

9. The chemical composition of water was one of the stumbling blocks to the problem of ascertaining the atomic weights of the elements. Although Cavendish seemed to have discovered the H_2O formula, Dalton apparently was

unaware of this result or rejected it while developing his atomic theory. As we shall see in Chapter 5, Avogadro came up with the correct formulation as well as the theoretical support for it.

The following site gives credit to Cavendish for proposing the formula for water (2.02 parts hydrogen to 1 part oxygen):

http://mattson.creighton.edu/History_Gas_Chemistry/Cavendish.html.

10. Available at http://www.archive.org/stream/lawsofgravitatio00mackrich/lawsofgravitatio00mackrich_djvu.txt (accessed January 6, 2011).

11. Ibid., Introduction to *Experiments to Determine the Density of the Earth.*

12. Available at http://arxiv.org/find (accessed January 27, 2011). This is the database home page; simply type in CODATA 2006 into the "Experimental full-text search" bar.

13. Ibid., 57.

CHAPTER 2

1. Available at http://www.elyrics.net/read/d/doors-lyrics/the-crystal-ship-lyrics.html (accessed January 17, 2011).

2. Available at http://www.elyrics.net/read/b/bob-seger-lyrics/night-moves-lyrics.html (accessed January 17, 2011).

3. Galileo Galilei, *Two New Sciences* (Madison: University of Wisconsin Press, 1974), 50.

4. S. Drake, "Galileo's First Telescopic Observations," *Journal of the History of Astronomy* 7 (1976): 157.

5. Available at http://en.wikipedia.org/wiki/Rømer's_determination_of_the_speed_of_light (accessed January 17, 2011).

6. Available at http://www-history.mcs.st-and.ac.uk/Biographies/Foucault.html (accessed January 17, 2011).

7. R. Staley, *Einstein's Generation: The Origins of the Relativity Revolution* (Chicago: University of Chicago Press, 2008), 212.

8. D. Livingston, *The Master of Light* (New York: Charles Scribner & Sons, 1973), 51.

9. Available at http://en.wikipedia.org/wiki/Double-slit_experiment (accessed January 17, 2011). This is one of the most important experiments in the history of science.

10. D. Livingston, *The Master of Light* (New York: Charles Scribner & Sons, 1973), 5.

11. T. Ferris, *Coming of Age in the Milky Way* (New York: Harper Perennial, 2003), 180.

CHAPTER 3

1. R. Boyle, "The Spring and Weight of the Air" in *The Works of Robert Boyle Vols. 1–7*, eds. E. Davis and M. Hunter (London: Pickering and Chatto,

1999). This is about as serious as serious scholarship gets—it seems to be sold out, but you can buy volumes 8–14 for only slightly less than $600. Or, you can find a discounted flight to the UK and visit the Whipple Museum of the History of Science, which has a section devoted to Boyle, for around the same price.

2. Available at http://en.wikipedia.org/wiki/Robert_Hooke (accessed January 9, 2011).

3. Robert Gunther, ed., *Early Science in Oxford*, privately printed. Referenced in Endnote 2 above.

4. M. Crosland, *Gay-Lussac: Scientist and Bourgeois* (Cambridge: Cambridge University Press, 1978), 7.

5. Available at http://www.chemistryexplained.com/Fe-Ge/Gay-Lussac-Joseph -Louis.html (accessed January 9, 2011).

6. Ibid.

7. For those who have taken a course in partial differential equations, this method of amalgamating the two laws may seem familiar. I got the idea from "separation of variables," which is a workhorse technique in partial differential equations.

8. Available at http://en.wikipedia.org/wiki/Equipartition_theorem (accessed January 9, 2011). This is one of the reasons Wikipedia is such a tremendous resource; when they asked for donations, I sent in a check. The site has a relatively simple explanation of the basic idea—and enough high-powered material to satisfy the most demanding theoretical physicist. In addition, there are some excellent photographs and graphics. When you click on the page, you get the article—but you might also look at the discussion section, which shows how the page evolved as part of WikiProjects Physics.

CHAPTER 4

1. Available at http://en.wikipedia.org/wiki/Absolute_zero (accessed January 17, 2011).

2. Ibid.

3. Available at http://en.wikipedia.org/wiki/Faraday (accessed January 17, 2011).

4. Available at http://en.wikipedia.org/wiki/P-V_diagram (accessed January 17, 2011).

5. Available at http://www.pbs.org/wgbh/nova/transcripts/3501_zero.html (accessed January 17, 2011). This is a transcript of a TV show from the justly celebrated PBS TV series *Nova*. The actual show lasts about 1:45 minutes; it's absolutely excellent, and you can see it just by clicking on the following link (also accessed January 17, 2011): http://www.cosmolearning.com/ documentaries/absolute-zero-2008/1/.

6. Available at http://thinkexist.com/quotes/damon_runyon/ (accessed January 17, 2011).

7. You can actually watch this happen on http://www.cosmolearning
.com/documentaries/absolute-zero-2008/1/ (accessed January 17, 2011).

8. Available at http://en.wikipedia.org/wiki/Bose–Einstein_condensate (accessed January 17, 2011).

9. T. Koshy, *Elementary Number Theory with Applications*, 2nd Ed. (Burlington, MA: Academic Press, 2007), 567.

10. Available at http://en.wikipedia.org/wiki/Bose–Einstein_condensate (accessed January 17, 2011).

11. Ibid.

12. This famous quote (it's so good I'll refer to it again in the chapter on Planck's constant) is somewhat controversial. Available at http://en.wikiquote
.org/wiki/Arthur_Stanley_Eddington (accessed January 17, 2011).

13. Yes, this is stunning—but true. See http://hypertextbook.com/facts/2007/
NadyaDillon.shtml (accessed January 17, 2011). There are several references to this as the price. If you want to compare this with the price of gasoline, though, there are 3.785 liters in a gallon, so liquid helium costs about $19 a gallon. Your favorite alcoholic tipple, assuming you are an individual of taste and discernment who does not indiscriminately purchase Two-Buck Chuck at Trader Joe's, costs more.

CHAPTER 5

1. Available at http://www.rationaloptimist.com/writings/cheer-life-only-gets
-better (accessed January 3, 2011).

2. Available at http://inventors.about.com/od/nstartinventions/a/nylon.htm (accessed January 2, 2011).

3. R. Feynman, *Six Easy Pieces* (New York: Basic Books, 1995), 4.

4. I'm continually amazed at what you can find on the Internet. Here's a translation of Avogadro's original paper; it is very easy to read—unlike Newton's *Principia*. Available at http://www.chem.elte.hu/departments/elmkem/
szalay/szalay_files/altkem/Avogadro_cikk.pdf (accessed January 2, 2011).

5. Available at http://en.wikipedia.org/wiki/Cannizzaro_reaction (accessed January 3, 2011).

6. E. J. Holmyard, *Masters of Chemistry* (Oxford: Oxford University Press, 1953), 257.

7. Available at http://en.wikipedia.org/wiki/Avogadro_constant (accessed January 3, 2011).

8. Available at http://www.usdebtclock.org/ (accessed January 3, 2011). This is a fascinating—and scary—website, because it continually updates all the factors that contribute to the national debt. When I accessed it, it read $13,939,520,000,000. By the time you read this, it will probably have ticked over $14 trillion.

CHAPTER 6

1. Available at http://en.wikipedia.org/wiki/William_Gilbert_(astronomer) (accessed January 5, 2011).

2. C. Gillmor, *Coulomb and the Evolution of Physics and Engineering in Eighteenth-Century France* (Princeton: Princeton University Press, 1971), 185.

3. Ibid., 207.

4. Ibid., 164.

5. Ibid., 198–210.

CHAPTER 7

1. Sooner or later, this had to come up. In the English system of units, which we use every day, pounds are a measure of weight, not mass. The difference between mass and weight is that mass doesn't change but weight does; our mass is the same here as on the Moon, but we weigh considerably less because the gravitational force acting on our mass on the Moon is much less than the gravitational force acting on our mass on Earth. It's the "little g" idea we encountered in the first chapter; little g is a local constant, big G a universal one.

Anyway, the unit of mass in the English system is the slug. I don't know where this comes from—possibly the word "sluggish" is related to it. Because we are so used to describing the weight of quantities in the English system with the word "pounds," it generally doesn't occur to us that this is not the same thing as mass. From $F = mg$, and the fact that g is roughly 32 feet per second per second at the Earth's surface, an object that weighs 32 pounds on Earth's surface has a mass of 1 slug.

One of the more confusing aspects of this is the common "misconversion" that 1 kilogram is equivalent to 2.2 pounds. I'm pretty sure where this comes from. A kilogram is defined as the mass of 1 liter of water; the liter is a measure of volume, and the gallon is also a measure of volume. One gallon is the same volume as 3.785 liters of water, and a gallon of water weighs about 8.35 pounds; 8.35 divided by 3.785 is 2.2, and so 1 liter of water weighs 2.2 pounds.

When doing calculations involving forces, it is easier to use the English system because the contribution from gravity is built into the definition of pound; it's actually a unit of weight, not mass. However, when doing calculations in the metric system, forces are measured in newtons; a newton is the result of accelerating a kilogram of mass at 1 meter per second per second. Consequently, when computing weights of objects in the metric system, masses in kilograms must be multiplied by the value of little g in the metric system (approximately 9.8 meters per second per second) to obtain their weight in newtons.

I'm amazed that this discrepancy hasn't been responsible for more computationally related disasters, but I'm not familiar with any.

2. A. Einstein, *The Evolution of Physics* (New York: Simon and Schuster, 1961), 44–47.

3. Available at http://www.chemteam.info/Chem-History/Rumford-1798 .html (accessed January 17, 2011). This is Rumford's original paper, and it is delightfully easy to read.

4. Ibid.

5. Available at http://en.wikipedia.org/wiki/James_Prescott_Joule (accessed January 17, 2011).

6. Ibid.

7. Available at http://www.archive.org/stream/reflectionsonmot00carnrich# page/n7/mode/2up (accessed January 17, 2011). This is a translation of Carnot's paper with a frontispiece picture of Carnot.

8. Available at http://en.wikipedia.org/wiki/Nicolas_Léonard_Sadi_Carnot (accessed January 17, 2011).

9. E. Broda, *Ludwig Boltzmann: Man, Physicist, Philosopher* (Woodbridge, CT: OxBow, 1983), 25. This delightful little book was one of my favorites. I was also intrigued to see it was published by a small publisher that reprinted another book that was tremendously helpful to me on another front. *The Bad Back Book*, by Jerry Wayne, is the true story of a Borscht Belt comedian with severe back problems. On a trip to India, he met a yogi who gave him a set of exercises that restored his back where all else had failed. I do those exercises to this day.

10. Ibid., 19.

11. Ibid., 11.

12. Available at http://www.poemhunter.com/poem/richard-cory/. I've always loved poetry—although I like it better when it rhymes and scans. Edward Arlington Robinson doesn't get the respect he deserves—nobody seems to know him nowadays—but Simon & Garfunkel composed a song about this poem that is well worth the three minutes of your time it takes to listen to it. It's available at http://www.youtube.com/watch?v=euuCiSY0qYs (both references accessed January 18, 2011).

13. Available at http://www.elyrics.net/read/j/jan-&-dean-lyrics/little-old -lady-from-pasadena-lyrics.html (accessed January 18, 2011).

CHAPTER 8

1. J. Heilbron, *The Dilemmas of an Upright Man* (Berkeley: University of California Press, 1986), 3.

2. Ibid.

3. Available at http://en.wikipedia.org/wiki/Philipp_von_Jolly (accessed January 11, 2011).

4. S. Brandt, *The Harvest of a Century* (Oxford: Oxford University Press, 2009) 29.

5. Available at http://en.wikipedia.org/wiki/Electromagnetic_spectrum (accessed January 11, 2011).

6. Available at http://en.wikipedia.org/wiki/Rayleigh-Jeans_Law (accessed January 11, 2011).

7. J. Bronowski, *The Ascent of Man* (Boston: Little, Brown, 1973), 336.

8. Available at http://www-groups.dcs.st-and.ac.uk/~history/Biographies/Taylor.html (accessed January 11, 2011).

9. J. Heilbron *The Dilemmas of an Upright Man* (Berkeley: University of California Press, 1986), 23.

10. Available at http://www.almaz.com/nobel/physics/1918a.html (accessed January 11, 2011).

11. R. Zimmerman, *An Amateur's Guide to Particle Physics: A Primer for the Lay Person* (Pittsburgh: Dorrance Publishing Co., 2003), 15.

12. A. Einstein, *Out of My Later Years* (New York: Citadel Press, 1976), 229.

13. Available at http://en.wikiquote.org/wiki/Arthur_Stanley_Eddington. Actually, WikiQuote says that this is misattributed to Eddington, and in fact is derived from a quote from the biologist J. B. S. Haldane, who originally said, "The universe is not only queerer than we suppose, it is queerer than we can suppose" in *Possible Worlds and Other Papers* (1927), 286. Everyone I know thinks it's an Eddington quote, though.

CHAPTER 9

1. Carl Sagan, *Cosmos* (New York: Random House, 1980), 134.

2. Available at http://www.le.ac.uk/ph/faulkes/web/stars/o_st_overview.html (accessed December 22, 2010).

3. Available at http://scienceworld.wolfram.com/physics/EddingtonLuminosity.html (accessed December 22, 2010).

4. Available at http://www.space.com/scienceastronomy/eso-massive-stars-discovered-100721.html (accessed December 22, 2010).

5. Available at http://www.unisci.com/stories/20022/0423021.htm (accessed December 23, 2010).

6. James Stein, *How Math Explains the World* (New York: HarperCollins, 2008), 196.

7. Kameshwar C. Wali, *Chandra: A Biography of S. Chandrasekhar* (Chicago: University of Chicago Press, 1990), 140.

8. Available at http://imagine.gsfc.nasa.gov/docs/science/know_l2/black_holes.html (accessed December 22, 2010).

CHAPTER 10

1. Available at http://www.cowboylyrics.com/lyrics/davis-skeeter/the-end-of-the-world-10980.html (accessed January 6, 2011).

2. Available at http://www.boardgamegeek.com/geeklist/30729/item/640692In the poem Taboo to Boot (accessed January 6, 2011).

3. Available at http://en.wikipedia.org/wiki/Alexander_Fleming (accessed January 6, 2011).

4. Available at http://www.quotationspage.com/quote/33774.html (accessed January 6, 2011).

5. Available at http://en.wikipedia.org/wiki/Crookes_tube (accessed January 6, 2011).

6. Available at http://en.wikipedia.org/wiki/Wilhelm_Röntgen (accessed January 6, 2011).

7. Ibid.

8. Available at http://nobelprize.org/nobel_prizes/physics/laureates/1901/perspectives.html (accessed January 6, 2011).

9. S. Brandt, *The Harvest of a Century* (Oxford: Oxford University Press, 2009), 80.

10. Available at http://nobelprize.org/nobel_prizes/peace/laureates/1962/# (accessed January 6, 2011).

11. Available at http://en.wikipedia.org/wiki/Crookes_tube (accessed January 6, 2011).

12. S. Brandt, *The Harvest of a Century* (Oxford: Oxford University Press, 2009), 41.

13. F. W. Dyson, A. S. Eddington, and C. Davidson, "A Determination of the Deflection of Light by the Sun's Gravitational Field, from Observations Made at the Total Eclipse of May 29, 1919," *Philosophical Transactions of the Royal Society of London*, Series A, Containing Papers of a Mathematical or Physical Character (1920): 332.

14. S. Brandt, *The Harvest of a Century* (Oxford: Oxford University Press, 2009), 258.

15. Ibid.

16. F. Levin, *Calibrating the Cosmos* (New York: Springer, 2006), 76–77.

17. S. Brandt, *The Harvest of a Century* (Oxford: Oxford University Press, 2009), 259.

CHAPTER 11

1. Like many scientists (or artists or businessmen), Bunsen was a workaholic; when absorbed in a problem, he refused to be disturbed. So much so, it is said, that on the day that he was to be married, he went to his laboratory and locked the door. Friends sent to fetch him were told to go away and let him work. You have to wonder if a story like that is apocryphal; but then, countless others have gotten cold feet on their wedding day. At any rate, there is no record of Bunsen ever having married.

2. S. Singh, *Big Bang: The Origins of the Universe* (New York: Harper-Collins, 2004), 237.

3. Available at http://www.pbs.org/wgbh/nova/transcripts/2311eins.html (accessed January 2, 2011).

4. Available at http://www.goodreads.com/author/quotes/10538.Carl_Sagan (accessed January 22, 2011).

5. K. C. Wali, *Chandra* (Chicago: University of Chicago Press, 1991), 6.

6. Available at http://nobelprize.org/nobel_prizes/physics/laureates/ (accessed January 2, 2011). This site has the complete list of prizes in physics, year by year. Clicking on each year will enable you to go deeper into the story, including the summary, the presentation speech, and the biographies of the winners.

7. Ibid.

8. K. C. Wali, *Chandra* (Chicago: University of Chicago Press, 1991), 62.

9. Available at http://www.ias.ac.in/jarch/jaa/15/115–116.pdf (accessed January 22, 2011).

10. Available at http://scienceworld.wolfram.com/physics/Chandrasekhar Limit.html (accessed January 22, 2011).

11. K. C. Wali, *Chandra* (Chicago: University of Chicago Press, 1991), 76.

12. Ibid.,125–126.

13. Ibid., 135.

14. Ibid., 138.

15. Ibid., 140.

16. Ibid., 10.

17. Ibid., 12.

CHAPTER 12

1. Available at http://www.jstor.org/pss/106639 (accessed January 25, 2011). JSTOR is one of the largest archives of scientific publications. However, anyone can get a peek at the first page by clicking on this link, and if you click on the "access options" link on this webpage, you should be able to see the whole paper for free (with the right connections to a participating institution), or you can purchase a download.

2. Ever since I read Thoreau's *Walden* in high school, I've been fascinated by how much money buys what in the different eras. Thoreau managed to exist on something like an expenditure of $28 for nine months. After writing the phrase "princely sum," I decided to see what $10.50 would buy in 1892. Meat was in the neighborhood of 7 to 10 cents a pound, bread was 5 cents a loaf. This indicates that prices have gone up in the range of forty to fifty times, so Leavitt's salary works out to somewhere around $25,000 per year. That puts her roughly at the midpoint of wage earners over twenty-five years old. The link http://en.wikipedia.org/wiki/Personal_income_in_the_United_States (accessed

January 23, 2011) has the following to say: "Of those individuals with income who were older than twenty-five years of age, over 42% had incomes below $25,000. . . . " Okay, not a princely sum (especially in Boston), but at least she wasn't being totally stiffed.

3. A. M. Lancaster, "The Discovery of the Sun Spots," *Appleton's Popular Science Monthly* September (1897), 683 (accessed January 25, 2011). In order to see this, use the search string "Holward Mira" in Google Books, and ignore repeated offers to search for "Howard Mira."

4. Available at http://en.wikipedia.org/wiki/Cepheid_variable (accessed January 25, 2011).

5. Available at http://en.wikipedia.org/wiki/Henrietta_Swan_Leavitt (accessed January 25, 2011). There's also a picture of Leavitt at her desk, in the starched white ruffled blouse we associate with her era.

6. H. Leavitt, *1777 Variables in the Magellanic Clouds*, Annals of Harvard College Observatory LX(IV) (1908), 87–110.

7. G. Johnson, *Miss Leavitt's Stars: The Untold Story of the Woman Who Discovered How to Measure the Universe* (New York: W.W. Norton & Company, 2005) 13.

8. G. Christianson, *Edwin Hubble: Mariner of the Nebulae* (New York: Farrar, Strauss, Giroux, 1995), 144.

9. Available at http://en.wikipedia.org/wiki/Henrietta_Swan_Leavitt (accessed January 25, 2011).

10. G. Christianson, *Edwin Hubble: Mariner of the Nebulae* (New York: Farrar, Strauss, Giroux, 1995), 141.

11. Available at http://www-groups.dcs.st-and.ac.uk/~history/Biographies/Doppler.html (accessed January 25, 2011).

12. Ibid.

13. J. Holberg, *Sirius: Brightest Diamond in the Sky* (Berlin: Praxis Publishing, 2006), 91.

CHAPTER 13

1. Available at http://en.wikipedia.org/wiki/History_of_general_relativity (accessed January 26, 2011).

2. R. Kirshner, *The Extravagant Universe* (Princeton: Princeton University Press, 2002), 56.

3. Available at http://www.physics.nyu.edu/faculty/sokal/weinberg.html (accessed January 26, 2011).

4. Available at http://mathworld.wolfram.com/EinsteinFieldEquations.html (accessed January 26, 2011).

5. Available at http://en.wikipedia.org/wiki/De_Sitter_ (accessed January 26, 2011).

6. G. Lemaître, "Un Univers homogène de masse constante et de rayon croissant rendant compte de la vitesse radiale des nébuleuses extra-galactiques," Annales de la Société Scientifique de Bruxelles, 47 (April 1927): 49.

7. A. Friedman "Über die Möglichkeit einer Welt mit konstanter negativer Krümmung des Raumes," *Zeitschrift für Physik* 21, no. 1 (1924): 326–332.

8. Available at http://en.wikipedia.org/wiki/Georges_Lemaître (accessed January 26, 2011).

9. Ibid.

10. Available at http://www.biblegateway.com/passage/?search=Genesis+1 -3&version=NIV (accessed January 26, 2011).

11. Available at http://en.wikipedia.org/wiki/Tired_light (accessed January 26, 2011).

12. Available at http://nedwww.ipac.caltech.edu/level5/Seitter/Seitter2_3_1 .html (accessed January 26, 2011).

13. Available at http://en.wikipedia.org/wiki/Steady_State_theory (accessed January 26, 2011).

14. Ibid.

15. Available at http://www.amnh.org/education/resources/rfl/web/essay books/cosmic/p_rubin.html (accessed January 26, 2011).

16. F. Levin, *Calibrating the Cosmos* (New York: Springer, 2006), 201.

17. M. Rees, *Just Six Numbers* (New York: Basic Books, 2000), 100.

18. Available at http://www.lbl.gov/supernova/ (accessed January 26, 2011).

19. Available at http://rationalwiki.org/wiki/Water_memory (accessed January 26, 2011). The moment this one came out, I knew it was hokum, and I couldn't believe it had actually made it into *Nature*, one of the premier journals of science.

20. Available at http://www.nuc.berkeley.edu/courses/classes/NE-24%20 Olander/cold_fusion.htm (accessed January 26, 2011).When I first read about this, I was extraordinarily hopeful. Pons and Fleischmann were respected scientists. But how could nuclear reactions occur from what seemed to be essentially chemical reactions? I crossed my fingers, because I believe that we will usher in a new Golden Age when—and if—fusion power becomes a practical economical method of producing energy. But, as time went on and no one could confirm the results, I was pretty certain that it just wasn't to be. However, this is the essence of what makes science unique among practically all other enterprises; truth must be independently reproducible. I can't claim that I knew it was bunk early, but after about two weeks, I would have bet strongly against it.

21. Available at http://www.crcnetbase.com/doi/abs/10.1201/9781420050 554.ch11 (accessed January 26, 2011). This one had a lot of believers, but to me it just seemed awfully unlikely. There were forces that operate only at extremely close distances, and forces that got stronger as you got further apart

(although it may be a little hard to believe physically, absence certainly makes the heart grow fonder)—but a force that operated exclusively over middle distances? And they didn't find it until the 1980s? Not bloody likely.

22. Available at http://www.astronomy.pomona.edu/Projects/moderncosmo/ Sean's%20mutliverse.html (accessed January 26, 2011).

23. Available at http://en.wikipedia.org/wiki/Goodbye,_Farewell_and_Amen (accessed January 26, 2011).

24. A. Christie, *Curtain* (New York: Berkley Books, 2000). I read it after I finished writing this book. I've never failed to enjoy an Agatha Christie, and while this wasn't the best of the Hercule Poirots, it was still a delight to read.

CODATA NOTE

1. Couldn't they have spent just a little time coming up with a more accurate acronym, or changing the name of the organization to the Committee on Data Accuracy for Technological Advancement, or something like that? Maybe they felt that acronyms are like loaves of bread: half an acronym is better than none.

2. Available at http://www.sing365.com/music/lyric.nsf/There's-No-Business -Like-Show-Business-lyrics-Irving-Berlin/ABA1A34F19D0E139482569700 00F2FA6 (accessed January 27, 2011).

3. Available at http://arxiv.org/find (accessed January 27, 2011). This is the database home page; simply type in CODATA 2006 into the "Experimental full-text search" bar.

4. Available at http://physics.nist.gov/cuu/Constants/index.html (accessed January 27, 2011).

INDEX